Urban Runoff Quality Management

WEF Manual of Practice No. 23

ASCE Manual and Report on Engineering Practice No. 87

Prepared by a Joint Task Force of the **Water Environment Federation** and the **American Society of Civil Engineers**

Larry A. Roesner, *Co-Chair*
Ben R. Urbonas, *Co-Chair* (ASCE)
William C. Pisano, *Vice-Chair*

John Aldrich
Geoff Brosseau
Carol Forrest
John P. Hartigan, Jr.
Bruce Hasbrouck
Edwin E. Herricks
Richard Horner
Wayne C. Huber
Jonathan E. Jones
G. Fred Lee
Eric H. Livingston
Wayne F. Lorenz
Dave Maunder
Lynn Mays
Peter E. Moffa
Lisann Chase Morris
Vladimir Novotny
Vincent J. Palumbo
Ronald A. Saikowski
James E. Scholl
W.J. Snodgrass
Malcolm R. Walker
William Whipple
Paul Wisner

Under the Direction of the
WEF Water Quality and Ecology Subcommittee of the **Technical Practice Committee** and
The Urban Water Resources Research Council of the American Society of Civil Engineers

1998

Water Environment Federation
601 Wythe Street
Alexandria, VA 22314–1994 USA
and
American Society of Civil Engineers
1801 Alexander Bell Drive
Reston, VA 20191–4400 USA

IMPORTANT NOTICE

The material presented in this publication has been prepared in accordance with generally recognized engineering principles and practices and is for general information only. This information should not be used without first securing competent advice with respect to its suitability for any general or specific application.

The contents of this publication are not intended to be a standard of the Water Environment Federation (WEF) or the American Society of Civil Engineers (ASCE) and are not intended for use as a reference in purchase specifications, contracts, regulations, statutes, or any other legal document.

No reference made in this publication to any specific method, product, process, or service constitutes or implies an endorsement, recommendation, or warranty thereof by WEF or ASCE.

WEF and ASCE make no representation or warranty of any kind, whether expressed or implied, concerning the accuracy, product, or process discussed in this publication and assume no liability.

Anyone using this information assumes all liability arising from such use, including but not limited to infringement of any patent or patents.

Library of Congress Cataloging-in-Publication Data

Urban runoff quality management / prepared by a joint task force of the Water Environment Federation and the American Society of Civil Engineers ; under the direction of the Water Quality and Ecology Subcommittee of the Technical Practice Committee and American Society of Civil Engineers.

p. cm. — (WEF manual of practice ; no. 23) (ASCE manual and report on engineering practice ; 87)

Includes bibliographical references and index.

ISBN 1-57278-039-8. — ISBN 0-7844-0174-8

1. Urban runoff—Management. 2. Water quality management. I. Water Environment Federation. II. American Society of Civil Engineers. III. Series: Manual of practice ; no. 23. IV. Series : ASCE manual and report on engineering practice ; no. 87.

TD201W337 1996

[TD657]

363.72′84—dc20 96-21244

CIP

ISBN 1-57278-039-8 and 0-7844-0174-8

Printed in the USA **1998**

Water Environment Federation

The Water Environment Federation is a not-for-profit technical educational organization that was founded in 1928. Its mission is to preserve and enhance the global water environment. Federation members are more than 42 000 water quality specialists from around the world, including environmental, civil, and chemical engineers, biologists, chemists, government officials, treatment plant managers and operators, laboratory technicians, college professors, researchers, students, and equipment manufacturers and distributors.

For information on membership, publications, and conferences contact

Water Environment Federation
601 Wythe Street
Alexandria, VA 22314-1994 USA
(703) 684-2400
http:\\www.wef.org

American Society of Civil Engineers

The American Society of Civil Engineers (ASCE) offers civil engineering professionals many opportunities for technical advancement, networking, and leadership and technical skill training. Also available to members are major savings on educational seminars, conferences, conventions, and publications. On the local level, chapters (called Sections and Branches) act as advocates in the public interest on local issues and present seminars and programs relevant to the needs of the local community. For more information, call 800-548-2723 (ASCE) or visit ASCE's web site, http://www.asce.org.

Manuals of Practice

(As developed by the Water Environment Federation)

The Water Environment Federation (WEF) Technical Practice Committee (formerly the Committee on Sewage and Industrial Wastes Practice of the Federation of Sewage and Industrial Wastes Associations) was created by the Federation Board of Control (now Board of Directors) on October 11, 1941. The primary function of the committee is to originate and produce, through appropriate subcommittees, special publications dealing with technical aspects of the broad interests of the Federation. These manuals are intended to provide background information through a review of technical practices and detailed procedures that research and experience have shown to be functional and practical.

ASCE Manuals and Reports on Engineering Practice

(As developed by the ASCE Technical Procedures Committee July 1930 and revised March 1935, February 1962, and April 1982)

A manual or report in this series consists of an orderly presentation of facts on a particular subject, supplemented by an analysis of limitations and applications of these facts. It contains information useful to the average engineer in his/her everyday work, rather than the findings that may be useful only occasionally or rarely. It is not in any sense a "standard," however, nor is it so elementary or so conclusive as to provide a "rule of thumb" for non-engineers.

Furthermore, material in this series, in distinction from a paper (which expresses only one person's observations or opinions), is the work of a committee or group selected to assemble and express information on a specific topic. As often as is practicable, the committee is under the general direction of one or more of the Technical Divisions and Councils, and the product evolved has been subjected to review by the Executive Committee of that Division or Council. As a step in the process of this review, proposed manuscripts are often brought before the members of the Technical Divisions and Councils for comment, which may serve as the basis for improvement. When published, each work shows the names of the committee by which it was compiled and indicates clearly the several processes through which it has passed in review so its merit may be definitely understood.

In February 1962 (and revised in April 1982), the Board of Direction voted to establish a series entitled "Manuals and Reports on Engineering Practice" to include the Manuals published and authorized to date, future Manuals of Professional Practice, and Reports on Engineering Practice. All such Manual or Report material of the Society would have been refereed in a manner approved by the Board Committee on Publications and would be bound, with applicable discussion, in books similar to past Manuals. Numbering would be consecutive and would be a continuation of present Manual numbers. In some cases of reports on joint committees, bypassing of Journal publications may be authorized.

American Society of Civil Engineers

Jerry L. Anderson, *Chair,* Water Resources Planning and Management Division Executive Committee, 1996

Jonathan Jones, *Chair,* Urban Water Resources Research Council

Authorized for publication by the Publications Committee of the American Society of Civil Engineers

James E. Davis, *Executive Director,* ASCE

Contents

List of Tables

List of Figures

Preface

The quality of urban stormwater was largely ignored in the design of urban drainage systems until approximately 1980. Previously, the focus was on efficient surface drainage and flood control, namely the effects of relatively large storm events. However, a number of engineers and scientists were becoming aware that runoff from the smaller, frequently occurring storm events was the cause of many observed negative effects in the nation's streams, lakes, estuaries, wetlands, and other receiving water systems downstream of, and within, the urban and urbanizing areas. Stream banks experienced accelerated erosion, stream habitat was degraded or lost, lakes and estuaries eutrophied at a faster rate, and the water quality in the receiving waters showed noticeable degradation during and sometimes after wet weather events.

Although many of the observed negative effects could also be attributed to causes other than stormwater runoff from urban areas (for example, nutrient- and sediment-laden runoff from agricultural areas, runoff from industrial areas, and unpermitted point sources), a general concern about the effects of uncontrolled urban growth on the nation's waterways began to take hold. In 1982, Florida became the first state to pass a law requiring that the urban runoff from the first inch of rainfall be treated to remove pollutants. Maryland and Delaware followed suit shortly thereafter. In 1988, the U.S. Environmental Protection Agency (U.S. EPA) promulgated its first draft regulations for the nationwide control of urban runoff quality and, in 1990, published final regulations governing municipalities with populations of more than 100 000 people.

Guidance for designing controls for urban runoff quality management has been published in the literature by many individuals; however, most of these publications are oriented toward a specific geographic region of the country, and guidance for application in other areas is not addressed. Also, many of these publications discuss the design of individual controls but not the subject of planning for urban runoff quality management on a watershed-wide basis.

This manual comprises a holistic view of urban runoff quality management. For the beginner, who has little previous exposure to urban runoff quality management, the manual covers the entire subject area from sources and effects of pollutants in urban runoff through the development of management plans and the design of controls. For the municipal stormwater management agency, guidance is given for developing a water quality management plan that takes into account receiving water use objectives, local climatology, regulation, financing and cost, and procedures for comparing various types of controls for suitability and cost effectiveness in a particular area. This guidance will also assist owners of large-scale urban development projects in cost-effectively and aesthetically integrating water quality control to the drainage plan. The manual is also directed to designers who desire a self-contained unit that discusses the design of specific quality controls for urban runoff.

Chapter 1 provides information on the sources and effects of chemical constituents in urban runoff. The chapter discusses chemical constituents found in urban runoff, with indications of the frequencies and concentrations in which these constituents are found. The effects of urban runoff on receiving waters are also addressed, including effects on water quality and ecology. The chapter also presents a section on environmental resource protection, which looks at urban runoff quality management from the more holistic receiving water management point of view, rather than the narrower and more traditional pollution minimization approach.

Chapter 2 is directed primarily at municipal agencies, who must develop and implement, or regulate, a plan of urban runoff quality management. The chapter addresses the development of such a plan from two perspectives: meeting U.S. EPA requirements of the 1990 regulations and achieving the objective of environmental resource management. Master-planning considerations take into account the type of receiving water, designated and desired beneficial uses, permitting requirements, financing, and integration of quality management with quantity management.

Chapter 3 addresses monitoring, bioassessment, and modeling. Information in this chapter supports the planning activities in Chapter 2 and the monitoring of control practices or devices that have been constructed.

Chapters 4 and 5 provide guidance on the design and implementation of source controls and treatment controls, respectively, for urban runoff quality management. Guidance is provided at the beginning of each chapter on how to evaluate tradeoffs between the various controls addressed. However, if guidance on the design of a specific control is all that is desired, the user may go directly to that section in the chapter and develop the design from the guidance given there. Familiarity with other parts of the manual is not required if the performance requirements of the control are known, or if a particular control is required by ordinance.

The user of this manual should understand that while established scientific practice underlies the treatment controls addressed here, considerable professional judgment is involved in the application. For this reason, the designer should be knowledgeable in aquatic chemistry and aquatic ecology—in addition to hydrology and hydraulics—or seek the advice of someone who is. Designers are advised to use their own common sense in applying the design criteria to a specific situation, combining theoretical concepts with experience in the joint application of the principles embodied in these four disciplines.

Finally, there is no way to separate quality management from quantity management of urban runoff. This manual deals with urban runoff quality;

the quantity aspects of urban runoff management are addressed only in a general way. For more details on quantity concerns, the reader is referred to *Design and Construction of Urban Stormwater Management Systems*, published by the American Society of Civil Engineers and the Water Environment Federation in 1992 (ASCE Manuals and Reports of Engineering Practice No. 77; WEF Manual of Practice No. FD-20), which is the companion manual on urban runoff quantity.

Principal Editors and Contributing Authors of this manual are

John Aldrich
Geoff Brosseau
Edwin E. Herricks
William C. Pisano
Larry A. Roesner
W.J. Snodgrass
Ben R. Urbonas

Authors of the manual also include

Carrol Forrest
John P. Hartigan, Jr.
Bruce Hasbrouck
Richard Horner
Wayne C. Huber
Jonathan E. Jones
Eric H. Livingston
Wayne F. Lorenz
Dave Maunder
Lynn Mays
Peter E. Moffa
Vladimir Novotny
James E. Scholl
Craig R. Smithgall
Malcolm R. Walker
William Whipple
Paul Wisner

In addition to the WEF Task Force and Technical Practice Committee Control Group members, reviewers and contributors include

Vince Berg
Doug Harrison
Jim Heaney
Angelo Musone

Robert Pitt
Earl Shaver
Paul J. Traina
Martin P. Wanielista

The final document was reviewed on behalf of ASCE by a review panel comprising the following members:

Richard Lanyon
Rooney Malcolm
Ronald Rossmiller

Authors' and reviewers' efforts were supported by the following organizations:

Camp, Dresser & McKee, Inc., Cambridge, Massachusetts (headquarters)
Environmental Engineering and Science, Seattle, Washington
Florida Department of Environmental Regulation, Tallahassee
Fresno Metropolitan Flood Control District, California
G. Fred Lee & Associates, El Macero, California
HDR Engineering, Inc., Tampa, Florida; Bellevue, Washington
Larry Walker Associates, Davis, California
Malcolm Pirnie, Phoenix, Arizona; White Plains, New York
Marquette University, Milwaukee, Wisconsin
McMaster University, Hamilton, Ontario, Canada
McNamee, Porter & Seeley, Ann Arbor, Michigan
Metropolitan Water Reclamation District of Greater Chicago, Illinois
Moffa & Associates, Syracuse, New York
Montgomery-Watson, Boston, Massachusetts
National Water Research Institute, Burlington, Ontario, Canada
North Carolina State University
Rhode Island Department of Transportation, Providence
Sitech Engineering Corporation, The Woodlands, Texas
University of Central Florida, Orlando
University of Colorado, Boulder
University of Florida, Gainesville
University of Illinois, Urbana
Urban Drainage and Flood Control District, Denver, Colorado
Wright Water Engineers, Inc., Denver, Colorado

Chapter 1
Introduction to Urban Runoff

MATERIAL ADDRESSED IN THIS MANUAL

URBAN RUNOFF EFFECTS AND CONTROL REQUIREMENTS. This manual focuses on the protection and enhancement of urban water resources through control of the transport of constituents into urban waterways by urban stormwater runoff. The manual emphasizes control of constituent discharges, reflecting the fact that chemical and particulate constituents in urban stormwater runoff play a key role in determining the negative effects of that runoff.

To provide a context for control requirements, receiving waters of potential concern are addressed, primarily in Chapters 2 and 3. Of these, streams tend to be the water body primarily affected by urbanization, and they are given special focus later in this introductory chapter. It is noted that the discussion of control addresses both traditional and developing approaches to control. Traditionally, control requirements have often been set without regard to the specific intended use of the receiving water. More recently, there have been attempts to improve the methods for measuring and assessing how constituents of stormwater runoff affect designated beneficial uses (see Chapters 1 and 3).

BEST MANAGEMENT PRACTICES. Best management practice, or BMP, as applied to urban runoff management was a term adopted in the 1970s to represent actions and practices that could be used to reduce the flow rates and the constituent concentrations in urban runoff. It reflects the developing and somewhat empirical engineering that applies to this area of practice and differentiates it from the more highly developed and predictable engineering associated with traditional wastewater treatment facility design.

A variety of urban BMPs are considered in Chapters 4 and 5. Source control practices and "passive" BMP systems (those not requiring active operational control or adjustment beyond routine maintenance) are described in some detail. Design considerations for "active" treatment technologies (those that are not passive, such as ultraviolet irradiation for bacteria [Craig and Tracy, 1993] or chemical precipitation for phosphorus) are not included in this manual. However, planning approaches, which might lead to the need for such systems, are discussed in Chapter 2. Design objectives for BMPs, and processes by which objectives can be established, are discussed in Chapter 2. It is noted that objectives can be stated in terms of technology (by specifying a particular control device) or in terms of quantitative effect (for example, by specifying a required degree of control or a maximum allowable effect). Because quantitative objectives can be defined for both hydrological parameters and constituent removal performance parameters, both of these are discussed. Examples of objectives based on hydrological parameters include peak flow rate and retention of a defined water volume for a specific period of time, while objectives based on chemical parameters include percent removal of specific chemical constituents and effluent concentration or mass discharge targets.

The understanding of the role of BMPs in reducing the potential toxicity of stormwater runoff is evolving. Data characterizing toxicological properties of urban runoff are sparse, and the scale-up of event-type phenomena to receiving water effects is not well understood. However, significant advances are being made in understanding this issue (see Chapters 1 and 3).

Best management practice technology is imperfect; for example, BMPs may not have a significant effect on the removal of soluble toxic substances. Therefore, pollution prevention (source control) and public education (also inherently a source control) are discussed in Chapter 4 as ways of providing some added improvement to effluent quality.

WATER QUALITY PARAMETERS. Water quality parameters addressed most in this manual are total suspended solids (TSS) and nutrients (nitrogen and phosphorus); this reflects current common practice in BMP design. In fact, TSS and nutrients are the primary constituents of stormwater runoff that can be controlled by the passive BMPs considered in this manual. It is noted that focus on these parameters is not a complete oversight of other parameters, because most other constituents of concern (for example, metals, hydrophobic organics) are reduced by the processes used to remove TSS, and most biochemically removable constituents will be reduced by processes that remove nutrients. Moreover, the two most widely documented effects of urban runoff on receiving waters are associated with sediment and nutrient enrichment.

Benefits do accompany the removal of other substances, such as metals (for example, copper, zinc, iron, and lead) if they are primarily particulate, hydrophobic organics, detritus, and bacteria. However, it is difficult to develop design criteria to control these substances because their removal depends on physiochemical factors that are intractable in design. All things considered, it is not unreasonable that present common practice focuses on particulate removal, and this approach is adopted in this manual.

It is noted that water quality parameters described in this manual typically use the total mass per unit volume (total concentration) as a basis for discussion. This is because the reference quantity used in many water quality standards is the total concentration. It is nevertheless recognized that the ecological significance of total concentration is the subject of much debate. This debate arises because in flowing water, the soluble form of substances (for example, in the case of copper) is the form that most directly causes toxicity to biota. However, the chemistry of the water, the substrate, and other factors make toxicity determination based on chemical speciation a complex process. This level of assessment is considered to be outside the present scope of practice targeted by this manual. References dealing with the details of aquatic toxicity may be sought in the literature.

DEFINITIONS OF TERMS

Constituents, pollutants, and contaminants are terms often used interchangeably in discussions of urban stormwater runoff. This section presents definitions for these and other commonly used terms in this manual. Definitions of various types of constituents of urban stormwater runoff are also presented. These latter definitions have been adapted from a technology-transfer publication of the University of Connecticut (1994):

- Site runoff coefficient—the ratio of direct runoff volume to rainfall volume, calculated over the duration of an event from beginning of rainfall to end of runoff resulting from that rainfall.
- Site imperviousness—the fraction of land surface that does not allow infiltration of rainfall at the start of a rainfall event.
- Constituent—a substance found in dissolved, colloidal, or particulate form in water that can be measured as a concentration.
- Pollutant—(a) a substance discharged at a rate that causes the receiving water ecosystem to become degraded or (b) a constituent in stormwater runoff that has a concentration and discharge rate (mass/time) that causes an impairment of designated beneficial uses of the receiving water. (The basis of a designated beneficial use [such as aesthetics, potability, recreational contact or noncontact, aquatic food consumption, or aquatic ecosystem protection] and its application in setting management objectives for urban stormwater runoff are presented in Chapter 2.)
- Contaminant—a term often used interchangeably with "pollutant" or to represent substances such as trace metals and synthetic organic

chemicals that have not historically been found in aquatic systems at present levels and that have been introduced by anthropogenic activities such as the processing of geological materials (such as lead from lead sulfide ores) or the release of new compounds not previously found in nature (such as pesticides).

- Water quality parameter—a physical, chemical, or biological characteristic, property, or representation of the quality of water. The parameter may be stated in qualitative terms (for example, an aesthetic property such as the presence or absence of trash) or in quantitative terms (for example, the concentration of a constituent in water).
- Pathogens—disease-causing microorganisms such as bacteria, viruses, and protozoan cysts that come from the fecal waste of humans and animals. Pathogens wash off the land from animal and pet waste. A more significant source of pathogens, especially in older urban areas, is the illegal connection of sanitary sewers to "separated storm sewers."
- Nutrients—substances that stimulate plant growth, such as nitrogen and phosphorus. Nutrients in polluted runoff can come from fertilizers, home lawn-care products, plant excretion, and yard and animal wastes. Other sources include cross connections and atmospheric inputs (for example, nitrous oxide emissions from automobiles).
- Sediments—sand, dirt, and gravel eroded by runoff that typically end up in stream beds, ponds, or shallow coastal areas. Poorly protected construction sites, roadways, suburban gardens and other unvegetated areas, and winter maintenance (sanding operations) can be significant sources of sediment. Soil erosion is more severe in areas under construction than in mature urban areas, which are stabilized by vegetation, pavement, houses, and sidewalks.
- Potentially toxic contaminants—substances that can harm the health of aquatic systems or human life. Toxins are created by a wide variety of human practices and products and include heavy metals (such as copper and lead), pesticides, and organic compounds. Oil, grease, and gasoline from roadways and some chemicals used in homes, gardens, and yards are toxic contaminants. An evolving area of knowledge is the role of automobile wear and tear (copper products, brake linings, oil drippings, and crankcase and radiator leaks), corrosion, and emissions and changes in vehicle technology and fuel constituents (for example, the removal of lead from gasoline) in contributing to these substances in urban runoff.
- Debris—trash, or solid waste, that often starts as street litter and is carried by runoff to waterways.

Of special note in further defining a pollutant are the terms "balanced" and "unbalanced," which have been used to describe oxygen concentrations in streams (Streeter and Phelps, 1925) and, in particular, eutrophication, which is associated with the excessive input of nutrients to the receiving water system. In the concept of balance, there are two significant *in situ* processes involved in a biological cycle: planktonic growth (that is, reduction of minerals and synthesis of organic materials) and bacterial degradation (that is, oxidation or decomposition of organic matter to liberate nutrients). When these two processes are in balance, the rate of eutrophication is minimal (Oswald and Golueke, 1966).

Therefore, pollution is a result of processes becoming out of balance, and the degree of pollution is a function of the degree of imbalance (for example, stimulation of planktonic growth or bacterially induced deoxygenation). Furthermore, concepts such as desired and designated beneficial uses have evolved to include other considerations such as human-health–related issues, consumptive uses, recreational uses, aesthetic values, and other water resource considerations.

STUDIES OF NONPOINT SOURCE POLLUTION AND REGULATORY BACKGROUND

The study of urban nonpoint source runoff in the U.S. has evolved from initial research into chemical constituents in dustfall and rainout conducted by the U.S. Geological Survey in the 1960s (Carter, 1961, and Leopold, 1968). The discovery that significant quantities of nutrients, pesticides, herbicides, and heavy metals were contained in urban runoff caused the U.S. Environmental Protection Agency (U.S. EPA) to require that regional urban planning agencies in the U.S. conduct planning studies regarding ways to reduce pollution from urbanized areas under Section 208 of the Clean Water Act. The studies, which came to be known as the "208 nonpoint source planning area studies," were conducted throughout significant metropolitan areas across the U.S. during the 1970s. The 208 studies were not successful because the profession had little experience in how to measure and quantify runoff quality or in the effectiveness of control measures in a dynamic system subject to the vagaries of urban hydrology. The unsuccessful 208 programs prompted U.S. EPA to fund the National Urban Runoff Program (NURP) of the early 1980s. While early studies aimed to define the magnitude of nonpoint source pollution relative to point sources and the significance of urban nonpoint sources (such as combined sewer overflows and separated storm drainage systems) relative to rural sources (such as agricultural land uses; wooded lands; and arid, sparsely vegetated lands), recent studies have focused on confirming the levels of pollution and effectiveness of pollution control efforts. A similar study path has been followed in Europe and Canada (Marsalek *et al.*, 1993), in which the focus on characterizing the relative importance of nonpoint sources has changed to the implementation of BMPs in agricultural and urban areas.

Best management practice, which has the roots of its definition in agricultural practice, refers to management practices and "soft engineering" methods to control pollutants in runoff. Agricultural management practices include leaving stubble in harvested fields to reduce wind erosion, fertilizer management to reduce costs, and crop rotation to increase yields; soft engineering controls include contour plowing, establishment of buffer zones between working fields and receiving waters, and check dams in drainage ditches to reduce sediment production. In an urban setting, BMPs are of two types: source controls (sometimes called "nonstructural" controls) and treatment controls (sometimes called "structural controls"). Source controls, which are described in detail in Chapter 4, are practices that keep chemical constituents from entering the runoff. Examples are covering chemical storage areas and/or diverting runoff away from such areas, street sweeping, and household hazardous waste recycling programs. Treatment control BMPs refer to devices that remove pollutants from the runoff. Examples are vegetated swales and buffers strips, infiltration, detention, and retention. Chapter 5 addresses treatment controls in detail.

After two decades of emphasis by U.S. EPA on the treatment of point source pollution as part of the Federal Clean Water Act (CWA) of 1972, remaining pollution problems now stem predominantly from diffuse and minor point sources. According to U.S. EPA, nonpoint source pollution represents more than half of the remaining water quality problems in the U.S. Through more recent legislation, U.S. EPA has developed a multifaceted approach to controlling runoff quality from these sources. Largely implemented through state agencies, this approach consists of overlapping programs covering general state strategies for nonpoint sources, municipal and industrial runoff quality, national estuary protection, protection of the U.S. Great Lakes, coastal zone protection, total maximum daily loads, and combined sewers. Runoff quality requirements have also indirectly been established through other federal and local environmental programs, such as environmental impact statements, endangered species programs, and wetlands preservation programs.

The CWA, as amended in 1987, and subsequent U.S. EPA regulations governing National Pollution Discharge Elimination System (NPDES) permits for stormwater discharges are the principal vehicles for controlling stormwater pollutants at the federal level. The 1987 amendments to CWA added Section 402(p), which established a framework for regulating municipal and industrial stormwater discharges as part of the NPDES program. In addition, in November 1990, U.S. EPA published regulations that established application requirements for stormwater NPDES permits.

U.S. EPA's application requirements for municipal discharges consist of two parts. Part 1 requires the discharger to collect existing information regarding stormwater dischargers, receiving waters, management programs, fiscal resources, and associated elements. In Part 2, a municipality is expected to take this information and formulate a stormwater management program designed to reduce the discharge of pollutants to the "maximum extent practicable."

Other laws/programs that directly or indirectly affect the control of stormwater pollutants are discussed in Chapter 2.

HYDROLOGICAL CHARACTERISTICS OF URBAN STORMWATER RUNOFF

HYDROLOGICAL CHANGES IN AN URBAN CATCHMENT. In general, urbanization can change a hydrologic cycle by

- Reducing the degree of infiltration and increasing the volume of runoff because of development of surfaces (changing slope, form, or cover);
- Changing the amount of depression storage because of regrading;
- Changing evapotranspiration because of removal of vegetative cover; and
- Reducing the travel time to a receiving body of water because of the construction of efficient sewer systems.

In extreme cases, small streams may be completely replaced by pipes and open channels after urbanization, resulting in streams and open channels that are completely dry between storms.

Imperviousness and Runoff. The relationship between imperviousness and runoff is worth considering in some detail. Figure 1.1 provides a good representation of the way the site runoff coefficient typically relates to site imperviousness. The figure was developed from more than 40 runoff monitoring sites throughout the U.S. (Schueler, 1994). The runoff coefficient (event runoff volume divided by event rainfall volume) ranges from 0 to 1, representing no runoff at one extreme and no infiltration at the other.

The first significance of the figure is that the runoff coefficient is closely correlated with the percentage of impervious cover, except at very low imperviousness. At low levels, other factors (soils and slopes) become important and imperviousness is a less perfect predictor of runoff coefficient.

It is interesting to note that the total runoff volume for a 100% paved parking lot is approximately 20 times that produced by an undeveloped meadow. This is indicative of the degree to which runoff volume can increase as sensitive lands are developed. Peak discharge, velocity, and time of concentration of stormwater runoff would also increase to a large degree in such a situation (see Table 1.1).

Another example of this is the calculated effect on a catchment's hydrol-

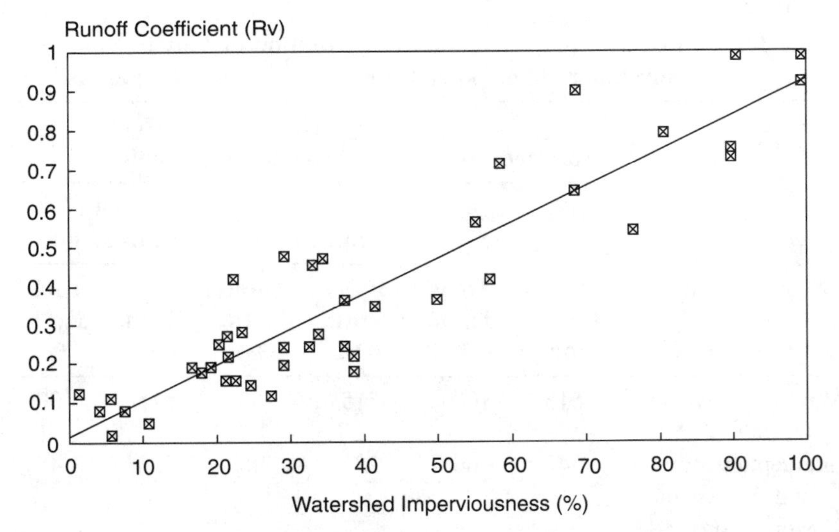

Figure 1.1 **Watershed imperviousness and the storm runoff coefficient.**

Table 1.1 **Comparison of 1 ac (0.4 ha) of parking lot versus 1 ac (0.4 ha) of meadow in good condition (from Schueler, T.R. [1994] *Watershed Protection Techniques: a Quarterly Bulletin on Urban Watershed Restoration and Protection Tools.* Center for Watershed Protection, Silver Spring, Md.,** 1, **1, with permission).**

Runoff or water quality parameter	Parking lot[a]	Meadow[b]
Curve number	98	58
Runoff coefficient	0.95	0.06
Time of concentration, minutes	4.8	14.4
Peak discharge rate, cfs,[a] 2/year, 24-hour storm	4.3	0.4
Peak discharge rate, cfs, 100-year storm	12.6	3.1
Runoff volume from 1-in. storm, cu ft[d]	3 450	218
Runoff velocity at 2-year storm, ft/sec[e]	8	1.8
Annual phosphorus load, lb/yr/ac[f]	2	0.50
Annual nitrogen load, lb/yr/ac	15.4	2.0
Annual zinc load, lb/yr/ac	0.30	—

[a] Parking lot is 100% impervious, with 3% slope; 200-ft (61-m) flow length; type 2 storm, 2-year, 24-hour storm = 3.1 in. (79 mm); 100-year storm = 8.9 in. (226 mm); hydraulic radius = 0.3; concrete channel; and suburban Washington ''C'' values.
[b] Meadow is 1% impervious, with 3% slopes, 200-ft (61-m) flow length, good vegetative condition, ''B'' soils, and earthen channel.
[c] cfs × 0.028 32 = m^3/s.
[d] cu ft × 0.028 32 = m^3.
[e] ft/sec × 0.304 8 = m/s.
[f] lb/yr/ac × 1.121 = kg/ha·a.

Table 1.2 **Effect of urbanization on distribution of May to November water budget for forested and urban areas.**[a]

Item	Forested areas		Urban areas with 40% impervious land			
			No infiltration		With infiltration	
	Depth, mm	Total depth, %	Depth, mm	Total depth, %	Depth, mm	Total depth, %
May to November rainfall	515	100	515	100	515	100
Interception storage and depression storage on impervious areas	342	66.5	235	45	235	45
Infiltration	155	30	100	20	200	40
Runoff	18	3.5	180	35	80	15

[a] Sandy soils assumed.

ogy because of a change in land use from a forested area to a typical single-family residential area. The results show a decrease in infiltration, depression storage, and interception storage (see Table 1.2). This results in a tenfold increase in the volume of runoff that reaches the receiving stream. Other observed changes in urban hydrology include the ratio of urbanized to nonurbanized peak flows. Data for three sites are provided in Table 1.3 (Urbanos and Roesner, 1992).

Table 1.3 **Ratio of peak runoff rates before and after development at three single-family residential sites.**

Return period, years	Post/preurbanization ratios of runoff peaks		
	New Jersey site[a]	Denver, Colorado, site[b]	Canberra, Australia, site[c]
2		57.0	9.0
10		3.10	4.7
15	3.0		
100		1.85	1.9

[a] 33-in. (840-mm) annual precipitation; based on modeling pre- and postdevelopment conditions using SCS TR-55 model and type II storm distribution.
[b] 15-in. (380-mm) annual precipitation; based on 8-year rainfall–runoff data record and 73-year simulation of pre- and postdevelopment conditions.
[c] 22-in. (550-mm) annual precipitation; based on statistical analysis of similar size adjacent developed and undeveloped tracts of land.

Effect of Imperviousness on Groundwater Recharge. As runoff volume increases with imperviousness, infiltration reduces. Groundwater recharge may be expected to reduce accordingly, which should, in turn, tend to cause lower dry weather stream flows. This is an effect that has been associated with urbanization; however, there are actual data that suggest this effect is not universal. An analysis of 16 North Carolina (U.S.) watershed areas could not find any statistical difference in low stream flow between urban and rural watersheds (Schueler, 1994). Nevertheless, other studies support the conclusion that the phenomenon exists. For example, dry weather flows dropped 20 to 85% after development in several urban watersheds in Long Island, New York (U.S.) (Schueler, 1994). And, in contrast, by evaluating low-flow statistics for urban streams in Toronto, Ontario (Canada), Belore (1991) found that low-flow rates (over a 7-day moving average and 10-year return period) increased during a 30-year urbanization period. One explanation offered for this counterintuitive result is that the additional water discharged to storm sewers from sources such as lawn irrigation and sanitary sewer exfiltration was sufficient to affect mass balances.

The net result of these studies is a mixed picture concerning the effects of urbanization on groundwater infiltration as inferred from base flow in streams. Therefore, it may be useful to critically evaluate these and other new studies case by case, paying particular attention to infiltration estimates distinct from base flow trends.

IMPERVIOUSNESS AS AN INTEGRATING CONCEPT. It is difficult to provide a measure of effect that simultaneously reconciles the needs of various scientists, engineers, and other stakeholders who must deal with the effects of urbanization. Nevertheless, some such common measure is needed if results are to be consistent when these individuals make decisions regarding individual development sites or the watershed as a whole.

One suggested index of effect is the amount of imperviousness on a given site or watershed. Imperviousness is the percent, or decimal fraction, of the total catchment covered by the sum of roads, parking lots, sidewalks, rooftops, and other impermeable surfaces of an urban landscape. Operationally, for mature urban areas, watershed imperviousness can be defined as the fraction of watershed area that is unvegetated.

Another way to describe imperviousness is that it represents the imprint of land development on a landscape, which, in simplistic terms, consists of two primary components: rooftops under which humans live, work, or shop and a transportation system consisting of roads, driveways, and parking lots. Increasingly, the "transportation system" often exceeds the rooftop component in terms of total impervious area created. For example, transportation-related imperviousness composed 63 to 70% of total impervious cover in 11 residential, multifamily, and commercial areas in the city of Olympia, Washington (U.S.) (Schueler, 1994, and City of Olympia, 1994).

Because imperviousness is a useful indicator for measuring the varying effects of land development on receiving waters and their aquatic systems, it can be viewed as providing a unifying theme in urban watershed protection (Schueler, 1994). It should be recognized that this is a convenient approximation only, not necessarily a realistic measure.

CONSTITUENTS IN URBAN STORMWATER RUNOFF

Two general approaches have developed in the literature for estimating contribution to chemical constituents in stormwater runoff: approaches based on human activities (such as pesticide application) and those based on general land-use categories (such as suburban downtown). Presently, land-use category approaches dominate the field of urban stormwater management to predict constituent concentrations. Primary land-use contributors are streets, roads, and highways; residential areas; commercial areas; industrial areas; and sites under development.

Historical case studies have provided alternative ways of estimating constituent loadings from urban areas or helped in developing methods for predicting pollutant levels based on urban characteristics (such as street accumulation or curb length). Implications for monitoring these studies are provided later in this chapter.

RESULTS FROM THE NATIONAL URBAN RUNOFF PROGRAM. National Urban Runoff Program research (U.S. EPA, 1983), which includes data from 28 urban sites throughout the U.S., is a compilation and evaluation of urban runoff data. The NURP study was funded and guided by U.S. EPA from 1978 to 1983, and although conducted at 28 local sites, it was reviewed, coordinated, and managed centrally.

One objective of the study was to characterize the water quality of discharges from separated storm sewers that drain residential, commercial, and light areas. The majority of samples collected in the study (see Table 1.4) were analyzed for eight conventional parameter sets. However, the study did not rigorously address issues such as the significance of elevated constituents' concentrations in urban runoff as they relate to the designated beneficial uses of receiving waters. This issue and the assessment/measurement of effects have been primary focuses of research and monitoring in similar studies of the 1990s.

The NURP study found that geographic location, land-use category, runoff volume, and other factors appeared to be of little use in explaining overall site-to-site or event-to-event variability. The NURP study determined that the best general characterization to predict characteristics of urban runoff at unmonitored sites is obtained by pooling site data from all sites (other than the open/nonurban ones). Pooled data for water quality characteristics from the NURP study are given in Table 1.4. In the absence of better

Table 1.4 **Overall water quality characteristics of urban runoff (U.S. EPA, 1983).**

Constituent	Typical coefficient of variation	Site median EMC[a]: For median urban site	Site median EMC[a]: For 90th percentile urban site
TSS, mg/L	1–2	100	300
BOD, mg/L	0.5–1	9	15
COD, mg/L	0.5–1	65	140
Total P, mg/L	0.5–1	0.33	0.70
Soluble P, mg/L	0.5–1	0.12	0.21
TKN, mg/L	0.5–1	1.50	3.30
NO_{2+3}–N, mg/L	0.5–1	0.68	1.75
Total Cu, μg/L	0.5–1	34	93
Total Pb, μg/L	0.5–1	144	350
Total Zn, μg/L	0.5–1	160	500

[a] Event near concentration.

information, the NURP study recommended the values given in Table 1.4 for planning purposes as the best description of the expected quality of urban runoff.

The event mean concentration (EMC) and the site mean concentration (SMC) are approaches to reporting the concentration of constituents in urban stormwater runoff and are used in Table 1.4 and subsequent chapters. Event mean concentration is the total mass of the constituent in the runoff event divided by the total volume of runoff during that event. Site mean concentration is the mean (from an arithmetic probability distribution) or median (from a log-normal distribution) value of all EMC values measured for a particular monitoring site.

In addition to assessing water quality characteristics, a portion of the NURP study also involved monitoring 120 "priority pollutants" in stormwater discharges (Thon, 1992). The study detected 77 priority pollutants in samples of stormwater discharges from the study sites, including 14 inorganic and 63 organic substances. Representative detection frequencies included lead (94%), zinc (94%), copper (91%), chromium (58%), arsenic (52%), pesticides (in particular, endosulfan, chlordane, and lindane) (15 to 19%), pentachlorophenol (19%), phthalates (22%), and fluoranthene (16%).

CONCENTRATIONS ESTABLISHED IN RECENT DATA SETS. **Representative Concentrations from One Urban Area**. The NURP database includes a limited number of water quality parameters. Where resources permitted, more parameters were monitored and SMC estimates were developed to determine the concentrations of other constituents (see Tables 1.5

Table 1.5 **Comparison of concentrations measured in Toronto waterfront studies with various water quality criteria (separated urban stormwater system discharge to Lake Ontario).**

Parameter		Discharge to sanitary sewer by-law target concentration	Discharge to storm sewer by-law target concentration	PWQO aquatic life (drinking water)[a]	Observed concentration dry weather outfall	Observed concentration wet weather outfalls
BOD	mg/L	300	15	—	7–19	420
Fecal coliforms	CNT/dL	300	15	100[b]	38 000–301 000	10 000–16E6
SS	mg/L	350	15	—	17–37	87–188
TP	mg/L	10	—	0.03	0.2–0.5	0.3–0.7
TKN	mg/L	100	—		1.8–4	1.9–3
Phenolics	mg/L	1	—	0.001	4–6	0.014–0.019
NO_3	mg/L	—	—	(10)	3.1–7.9	1.1–2.1
Al	mg/L	50	—	—	0.25–0.35	1.2–2.5
Fe	mg/L	50	—	0.3	0.63–1.0	2.7–7.2

Cr	mg/L	5	0.2	0.1	0.008–0.13	0.009–0.025
Pb	mg/L	5	0.05	0.025	0.008–0.012	0.038–0.055
Mn	mg/L	5	—	—	0.11–0.17	−0.12–0.17
Se	mg/L	5	—	0.1	<0.001	<0.001
Ag	mg/L	5	—	0.000 1	<0.01	0.002–0.005
Cu	mg/L	3	0.01	0.005	0.040–0.071	0.045–0.46
Ni	mg/L	3	0.05	0.025	0.008–0.012	0.009–0.016
Zn	mg/L	3	0.05	0.030	0.42–0.065	0.14–0.26
Total cyanide	mg/L	2	—	0.005	—	0.005
As	mg/L	1	—	0.1	0.002–0.004	<0.001
Cd	mg/L	1	0.001	0.000 2	<0.002	0.001–0.024
Hg	mg/L	0.1	0.001	0.000 2	<0.000 01	0.000 04–0.000 06
PCBs	μg/L	0	0	0.001	<0.02	
Solvent extractable	—	—	—	—	5–11	—

[a] Guideline values in brackets are for drinking water.
[b] Guideline for swimming.

Table 1.6 Comparison of concentrations (μg/L) measured in Toronto waterfront studies with guidelines for organic parameters (separated urban stormwater system discharge to Lake Ontario).

Compound, μg/L	Guidelines	Observed concentration, dry weather outfalls	Observed concentration, wet weather outfalls
Phenols	2.0	8	17
Toluene	300	0.02	—
Benzene	300	0.02	—
α-BHC[a]	0.092	0.001	0.001
γ-BHC[a]	0.186	0.000 5	0.001
Total PCB[b]	0.001	<0.005	<0.25
Anthracene	—	<0.02	0.061
Fluoranthene	42	<0.02	0.782
Pyrene	—	<0.02	0.615
Benzo(A)anthracene	—	<0.04	0.249
Chrysene	—	<0.02	0.333
Hexachlorobutadiene	0.1	<0.000 4	0.000 24
Bis-2-ethyl hexyl phthalate	6	7.4	—
Dichlorobenze 1,2	2.5	<0.02	—
Dichlorobenze 1,3	—	<0.02	—
Dichlorobenze 1,4	4.0	<0.02	—
Trichlorobenzene 1,2,4	0.5	0.002	0.005
Trichlorobenzene 1,2,3	0.9	<0.000 1	0.002
Trichlorobenzene 1,3,5	0.65	<0.000 05	<0.000 4
Tetrachlorobenzene 1,2,3,4	0.1	<0.000 05	<0.000 4
Pentachlorobenzene	0.03	<0.000 05	0.000 8
Hexachlorobenzene	0.006 5	<0.000 05	0.000 3
Heptachlor epoxides and heptachlor	0.01	<0.000 01	<0.000 05

[a] Benzenehexachloride.
[b] Polychlorinated biphenyl.

and 1.6 and papers published in Torno *et al.* [Eds.], 1994, for recent data). These data sets provide several advantages. A data set from the same site improves the assessment of water quality because it is often difficult to find an internally consistent set of data that includes most parameters. New statistical techniques for detection-limit data now permit estimation of the SMC, even if 80% of the EMC data from a site are below analytical detection limits. Moreover, new analytical equipment and improved sampling and analytical methodologies have resulted in lower detection limits and more rigorous quality assurance/quality control.

Concentrations measured in separated urban storm sewers for dry weather and wet weather conditions are given in Tables 1.5 and 1.6 and are compared to various water quality targets. Catchments that are drained by these storm sewers serve low-density suburban areas (that is, drainage areas ranging from

10 to 100 ha) of Metropolitan Toronto along the Toronto waterfront (D'Andrea *et al.*, 1993). For many conventional parameters and trace metals, observed concentrations (see Table 1.5) exceeded relevant receiving water standards and even targets established for discharges to storm drains. For a variety of other compounds (see Table 1.6), either urban source concentrations were below water quality targets or water quality targets were not established.

In general, these and similar literature data sets (Cooke *et al.*, 1995) can be used to define exclusionary criteria for parameter selection, provided that additional criteria, such as a knowledge of what constituents have the potential to be in the water, are used to define the parameters to be excluded.

GENERAL LITERATURE. The following is a list of references for individual sources of constituents in urban stormwater runoff:

- Atmospheric deposition, general levels (Halverson et al., 1982; Harrison and Johnston, 1985; Ng, 1987; and Novotny and Chesters, 1981).
- Atmospheric deposition by land-use type (Novotny and Kincaid, 1982; Pitt and Barron, 1989; and Randall *et al.*, 1982), geology (Pitt, 1979), wet fall–dry fall (Bannerman *et al.*, 1984), and dry deposition (Department of City Development, 1981).
- Street refuse deposition (see U.S. EPA, 1984, for particles greater than 60 μm).
- Vegetation (Halverson *et al.*, 1984, and Heaney and Huber, 1973).
- Traffic (Harrison and Johnston, 1985; Pitt, 1979; Sartor and Boyd, 1972; Shaheen, 1975; and Strecker *et al.*, 1987).
- Deicing chemicals, when applied (Field *et al.*, 1974; Lord, 1988; Oberts, 1986; and Zariello, 1990).
- Impervious and pervious surfaces:
    - Accumulation on impervious surfaces—buildup (Heaney and Huber, 1973; HEC, 1975; Huber, 1986; James and Boregowda, 1986; U.S. EPA, 1971; and Whipple *et al.*, 1978);
    - Retrainment from impervious surfaces (Cowherd *et al.*, 1977, and Pitt, 1979);
    - Winter accumulation (Bannerman *et al.*, 1984);
    - Washoff (Sartor and Boyd, 1972; Sartor *et al.*, 1974; and Zison, 1980); and
    - Losses with snow melt processes (Bengtsson, 1982; McComas *et al.*, 1976; Novotny, 1987 and 1988; Oberts, 1986 and 1990; and Westerström, 1990).
- Cross connections and illicit discharges (Pitt *et al.*, 1990a and 1990b, and Schmidt and Spencer, 1986).
- Grain size distribution (Sartor et al., 1974).

The following is a list of references for other substances in urban runoff (see Marsalek and Torno [Eds.], 1993; Torno *et al.* [Eds.], 1994; and historical references cited below):

- Microorganism (Glenne, 1984, and Olivieri *et al.*, 1977 and 1989);
- Pesticides and herbicides (U.S. EPA, 1984);
- Plasticizers (U.S. EPA, 1984);
- Polychlorinated biphenyls (Marsalek, 1986);
- Petroleum hydrocarbons (Hoffman, 1985, and Hoffman *et al.*, 1984);
- Polyaromatic hydrocarbons (Fam *et al.*, 1987; Forster, 1990; Hoffman *et al.*, 1984; and Marsalek, 1990);
- Toxicity (Pitt and Barron, 1989); and
- Atmospheric acidity and its influences (Forster, 1990, and Novotny and Kincaid, 1982).

Measurement of Other Chemicals. There are more than 60 000 chemicals used in the U.S. In relative terms, only a few are analyzed in a typical extended suite of analysis (approximately 100 to 150) or are regulated in terms of having receiving water standards. More recent studies (Waller *et al.*, 1995) have identified pesticides such as diazinon, which is a toxicant that is not typically tested for in water quality monitoring programs, in runoff in the southern U.S. Therefore, a designer should be encouraged to consider potential unknown factors about chemicals used in watersheds to establish a water quality management or monitoring program.

CONCENTRATIONS IN RUNOFF FROM OTHER STORMWATER SOURCES. Combined Sewer Overflows. Representative constituent levels for combined sewer overflows (CSOs) based on late 1970s monitoring have been established by U.S. EPA (1978). Although CSOs have been scrutinized by researchers and U.S. EPA and a large number of data sets on their use have been compiled, a statistical analysis comparable to NURP studies is only now evolving. Therefore, only ranges and mathematical averages have historically been reported. Some analysis has emphasized EMCs for CSOs (Driscoll and James, 1987).

The significance of CSOs, relative to secondary wastewater treatment plant (WWTP) discharges to surface waterways, was demonstrated in a U.S. EPA nationwide assessment conducted in 1978. For example, for the same land area, loadings from CSOs relative to loadings from secondary WWTPs had the following relationships: biochemical oxygen demand (BOD), equal; lead and suspended solids (SS), CSOs were 15 times larger; and total nitrogen and phosphorus, CSOs were one-quarter to one-seventh. The relative significance of CSOs and separated storm sewers in a watershed is site specific. For initial screening for planning purposes on a watershed basis, consideration of the relative volumes of runoff from different types of land areas (for example, separated storm drains, highways, or CSOs) and EMCs for these land areas is an appropriate approach. Where data for CSOs are not available, an EMC can be estimated from the relative volume of stormwater flow and sanitary sewer flow in the "overflow event" and the stormwater EMC and sanitary wastewater concentration (estimated as the dry weather flow concentration or the average wet weather concentration in the WWTP inflow).

Industrial Sources. Data for concentrations in runoff from industrial sources are evolving under the NPDES program. A designer is encouraged to contact local agencies and trade associations for relevant data.

Highway Sources. The U.S. Federal Highway Administration (FHWA) maintains a national database for highways similar to the NURP database. In the FHWA database, sites were sampled during the same time period (early 1980s) as the NURP database and were summarized by Driscoll *et al.* (1990) using similar statistical techniques (such as EMCs and reduction to approximately log-normal distributions).

The FHWA database contains concentration data (see Watershed Planning section) for 10 water quality parameters. The database has been summarized for urban areas and rural areas for two highway traffic densities: greater than and less than 30 000 average daily traffic (ADT). (Traffic density is the total number of vehicles that drive past a specific point in both directions in all lanes expressed on a daily basis.) Investigators such as Strecker *et al.* (1990) have provided probabilistic representations of the data (see Watershed Planning section). Recent synthesis documents (for example, Young *et al.,* 1996) stress that the categories of "urban/rural" and "less than/greater than" continue to be the accepted approach for estimating pollutant concentrations in highway runoff. However, some question still exists about the use of the database—most of the urban sites had an ADT of greater than 30 000, whereas most of the rural sites had an ADT of less than 30 000. This led Young *et al.* (1996) to summarize the data for two categories, urban areas (ADT $>$ 30 000) and rural areas (ADT $<$ 30 000), and to not provide data for the opposite categories of urban areas (ADT $<$ 30 000) and rural areas (ADT $>$ 30 000). Other studies that focus on data from only one site (such as Kerri *et al.,* 1985; Barrett *et al.,* 1995; and Thomson *et al.,* 1995 and 1997) have found additional predictive factors for constituent concentrations, such as ADT during the storm and interevent periods.

In the absence of site-specific data, these values are useful for planning and other purposes.

LESSONS LEARNED FROM PREVIOUS WATER QUALITY MONITORING

STATISTICAL CONCEPTS FOR URBAN RUNOFF WATER QUALITY EVALUATION. The NURP studies focused on evaluating EMCs. In most cases in these studies, the total load from the runoff event was more important than the individual concentrations within the event because the nature of the effect (for example, loadings to a receiving water) on the receiving water did not occur instantaneously but, rather, during an extended period of time.

Evaluation of the NURP database (U.S. EPA, 1983) and European data (Harremöes, 1988) revealed that the probability distribution of EMCs followed a probability distribution reasonably represented as log-normal. Additional studies (Driscoll and James, 1987) have also demonstrated that concentration data for other sources, such as CSOs, WWTP influent and effluent streams, and agricultural runoff, can be taken to have log-normal distributions.

The representation of the fundamental probability distribution of EMCs as log normal has a number of benefits (U.S. EPA, 1983, and Marsalek, 1991), including the following:

- Concise summaries of highly variable data can be developed;
- Comparisons of results from different sites and events are convenient and are more easily understood;
- Conclusions can readily be made concerning frequency of occurrence (one can express how often values exceed various magnitudes of interest);
- A more useful method of reporting data than the use of ranges is provided (one that is less subject to misinterpretations);
- A framework is provided for examining "transferability" of data in a quantitative manner;
- Data below the detection limit can be extrapolated; and
- Loadings can be obtained by multiplying the EMC by the total volume of runoff.

Another important statistical concept is the analysis of detection-limit data from knowledge of the probability distribution function. For many parameters, 20 to 80% of the sample values from a sample set are below detection limits. This can pose a problem in determining the true population mean and variance. An effective approach is to use a probability distribution estimation technique for this left-censored data set.

The variance structure of the term "loadings" also influences the method for reporting water quality data. To allow comparison of data between sites, investigations of the 1960s and 1970s reported data using units of unit loads (loadings per unit area of catchment). This is in contrast to recent data, which emphasize concentrations. Because unit loads include variance (because of both hydrological and concentration components), the concentration approach is emphasized in this manual.

PREDICTIVE RELATIONSHIPS OF URBAN RUNOFF QUALITY. Several potential predictors of constituent concentrations in runoff were evaluated in the NURP studies (U.S. EPA, 1983). The results are summarized in the following sections.

Land-Use Effects. The nationwide analysis in the NURP studies did not detect a significant statistical correlation of EMCs to the geographical locations of the sites studied throughout the U.S. In addition, three typical land uses, "residential, mixed, and commercial," were not found to be statistically different. Only open land uses and nonurban land uses were found to be significantly different (using statistical criteria) from these three land uses.

Runoff Volume Effects. A total of 67 sites from 20 of the NURP projects were examined for possible correlation between volume and EMC for nine constituents. The NURP study concluded that there is no significant linear correlation between EMCs and runoff volume.

Other Watershed Factor Effects. Factors such as slope, soil types, and rainfall characteristics are all potentially important. However, in a statistical sense, these factors did not have any consistent significance in explaining observed similarities or differences among individual sites.

IMPLICATIONS FOR MONITORING PROGRAMS. Monitoring programs for characterizing the quality of stormwater runoff, defining the effectiveness of BMPs, and developing relationships between runoff quality and various factors require increasingly onerous amounts of data and numbers of monitoring locations. Therefore, results from existing monitoring programs can be used to provide reasonably clear direction for typical stormwater quality monitoring issues in the 1990s. These issues include

- Defining differences in the concentration of stormwater chemical constituents over time at a site (for example, lead concentrations have declined because of their removal from gasoline);
- Defining a structured approach to measuring site-specific environmental effects and the implications of stormwater toxicity measurements (Herricks *et al.,* 1994);
- Applying scale-dependent factors (such as time and space) to a monitoring program design (for example, in a watershed in which BMPs have been completely implemented); and
- Defining the amount and type of monitoring data required to show significant differences between inflow and outfall loadings from a BMP facility (for example, one might establish 20% differences with an 80% level of confidence [a minimal statistical test] in a monitoring program).

Moreover, the issue of monitoring data to show differences between inflow and outfall of a BMP can be further addressed by analyzing NURP data, data from specific sites (see Torno *et al.* [Ed.] [1994] for examples of urban sites, and Thomson *et al.* [1995] for an example of a highway site), and data collected from a study of 18 sites in Austin, Texas (U.S.) (Cheng *et al.*, 1994).

Stormwater quality monitoring is sufficiently well developed today that direction can be given to new monitoring programs. Monitoring the environmental effects of stormwater runoff is a combination of art and science and requires hypothesis testing, evaluation of what "effects" are measurable, and an adaptive approach to program design (Hollings [Ed.], 1978).

National Urban Runoff Program research (U.S. EPA, 1983) and the FHWA database (Driscoll *et al.*, 1990), which were gathered in the early 1980s, are still considered the definitive studies available to characterize the concentration of substances such as SS, nutrients, metals, and oxygen-demanding substances (for example, chemical oxygen demand and BOD).

Recent NPDES monitoring data typically are consistent with these constituent concentrations except for lead, which displays lower levels because of its removal from gasoline. However, more recent data should be used to characterize many parameters, such as synthetic organic chemicals and polynuclear aromatic hydrocarbons (PAHs), because of the use of lower detection limits and better quality assurance/quality control. In addition, because of different levels of detection limits for parameters such as PAHs, some of these programs have varying degrees of utility.

As shown in NURP data, concentration (EMC) and runoff volume are independent (such that there is no significant correlation). The influence of other factors, such as interevent period and event duration, may be influential at a particular site (Barrett *et al.*, 1995, and Thomson *et al.,* 1997).

Studies have suggested that a minimum of 15 to 20 (Thomson *et al.*, 1995) or 18 to 30 (Cheng *et al.*, 1994) samples per site are required to provide an unbiased estimate of an SMC. In earlier studies, such as one conducted in Austin, Texas, approximately 40 monitoring sites were thought to be needed to show differences between two types of land uses and three sizes of watersheds to obtain four to eight mean concentration values for each combination of land use and watershed scales (Cheng *et al.*, 1994).

Furthermore, a practical measurement program can be used to specify the minimum number of flow-proportioned samples necessary to obtain a reasonable estimate of SMC, where the minimum number is a compromise between the number of samples suggested by a thorough statistical analysis and economic and resource considerations. For example, for sites typical of the temperate climates of the northeastern U.S., a reasonable minimum number would be 8 to 10 events, while a larger number of events may be needed for extremely left-censored data. For such a minimum number, obtaining two to three samples per site to address site-specific questions is not useful. Aggregation of two to three samples per site and data from several sites in an urban area may provide useful data for establishing SMCs; however, the additional variance introduced by this process should be evaluated.

Additionally, although first-flush phenomena have been observed in relatively small catchments (see Thomson *et al.*, 1995, for a Minnesota highway example), the existence of first-flush phenomena in large catchments is less clear.

Finally, performance assessment data from existing BMP databases can be used to define the amount of data required to evaluate the performance of new BMPs. For example, for a wet detention pond BMP (see Chapter 5), the following performance (removal of constituents) might be observed: SS, 70%; lead, 70%; total phosphorus, 50%; and total zinc, 20%. Compilations of performance-monitoring data (Schueler *et al.*, 1992) are rapidly accumulating; one method for decreasing the range of observed performance (for example 30 to 90% for TSS) is the use of standardized BMP data reporting protocols (Urbonas, 1994).

Clear monitoring guidance cannot be given for several biological and ecological properties of stormwater. The database characterizing the toxicological properties of urban runoff is sparse, and the scale-up of event-type phenomena to receiving water effects is not well known (see Herricks *et al.*, 1994). The role of BMPs in minimizing or reducing the potential toxicity of stormwater runoff is an evolving science. However, significant advances are currently being made in developing an understanding of this issue, as discussed at the conclusion of this chapter and in Chapter 3.

EFFECTS OF URBANIZATION ON RECEIVING ENVIRONMENTS

Urbanization affects all components of the environment, such as air quality, surface water, groundwater, soil quality, and the habitats for animals, including humans. The quality of the existing local environment, which may range from pristine to degraded, is influenced by the form and characteristics of the existing development within the watershed. New development, redevelopment, and retrofitting provide opportunities for changing the characteristics of the receiving environment. The form of new development, together with the existing quality of the local environment, determines whether these projects will cause further degradation or improve the local environment.

The alteration of the hydrological cycle in an urban watershed was described earlier in this chapter. This section describes the effects of these alterations on receiving environments. Effects from the alteration of the hydrological cycle are presented in an integrated way to assist designers in developing appropriate solutions for specific conditions. This section discusses effects on streams, wetlands, lakes, groundwater, and the biological habitats supported by these aquatic systems. Emphasis is given to streams because the relative distribution of surface waters affected by urban areas has been estimated by Herricks (1991) as follows:

- Receiving streams and rivers: 85%. More than 80% had an average annual flow less than 8 500 L/s (300 cfs), with many having a flow of less than 8.5 L/s (0.3 cfs).

- Receiving lakes: 5%.
- Small ponds, shallow backwater areas: less than 0.1%.
- Estuaries and oceans: 10%.

Not captured in this survey is the large number of headwater streams affected by urbanization.

EFFECTS OF STORMWATER RUNOFF ON STREAM ECOSYSTEMS. Effects in urban streams can be loosely grouped into four categories: changes to stream hydrology, stream form, water quality, and aquatic ecology (Schueler, 1992). The extent of an alteration is a function of the climatic regime (wet or dry) and change in land use.

The example presented in this section for streams is based on the urbanization of a forested headwater watershed in a relatively wet, temperate area on the eastern seaboard of the U.S. Therefore, a designer should consider the hydrologic characteristics of other climatic areas and an individual stream's morphological setting to assess the potential effect of hydrological alterations on particular receiving streams.

Changes in Stream Hydrology. The net effect of conventional development practices on an urban stream is a dramatic change in the hydrologic regime of the stream. Effects include

- An increase in the magnitude and frequency of runoff events of all sizes;
- Delivery of more of the stream's annual flow as surface storm runoff rather than base flow or interflow; and
- Increases in velocity of flow during storms.

Changes in Urban Stream Morphology. Stream channels in urban areas respond and adjust to the altered hydrologic regime that accompanies urbanization. The severity and extent of stream adjustment is a function of the degree of watershed imperviousness. Examples of stream adjustments and their consequences include

- Increased stream cross-sectional area to accommodate higher flows;
- Significant downcutting of the stream channel (unless the bottom is heavily armoured), where widening is prevented by road or pipeline crossings;
- Increased sediment loads in the stream because of increased stream bank erosion and upland construction site runoff;
- Modification of the streambed (typically, the grain size of channel sediments shifts from coarse-grained particles to a mixture of fine- and coarse-grained particles); and
- Modification of the stream through straightening and/or lining by humans to "improve" drainage and reduce flooding risks in intensively

urbanized areas (headwater streams tend to suffer disproportionately from enclosure).

Additionally, stream crossings by roads and pipelines change such stream plan-form characteristics as location and meander pattern. These structures may be heavily armored to withstand the downcutting power of stormwater.

A critical issue is the level of development at which stream morphology begins to change significantly. Research models developed in the Pacific Northwest (U.S.) suggest that a threshold for urban stream stability exists at approximately 10% imperviousness of a watershed (see Figure 1.2) (Booth, 1991, and Booth and Reinelt, 1993). Watershed development beyond this threshold consistently results in unstable and eroding channels. The rate and severity of channel instability appears to be a function of subbankful floods, the frequency of which can increase by a factor of 10 even at relatively low levels of imperviousness (Hollis, 1975; Macrae and Marsalek, 1992; and Schueler, 1994).

Changes in Stream Water Quality. Changes in stream water quality are associated with two phases of urbanization. During the initial phase of development, an urban stream can receive a significant pulse of sediment eroded from upland construction sites, even if erosion and sediment controls are

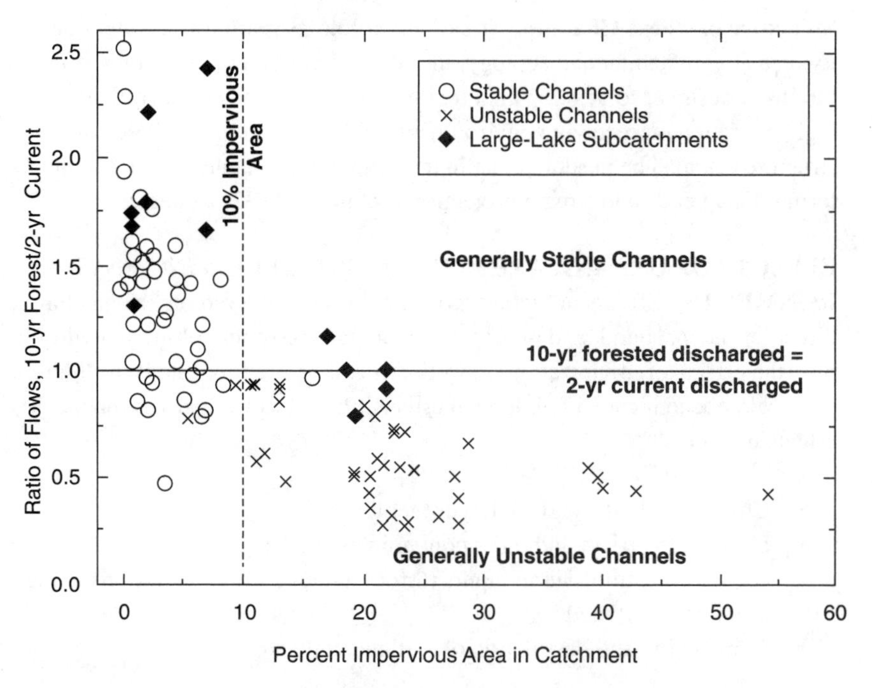

Figure 1.2 **Channel stability as a function of imperviousness (Booth and Reinelt, 1993).**

used. Sediment contributions from the land surface typically decline to less than predevelopment contributions after upland development stabilizes and is replaced by increased stream-bank erosion. In the second phase of urbanization, the dominant source is the washing off of accumulated deposits from impervious areas during storms (Schueler, 1992).

In general, constituent concentrations in urban streams are one to two orders of magnitude greater than those reported in forested watersheds (see earlier section, Constituents in Urban Stormwater Runoff). Their degree of loading has been shown to be a direct function of the percentage of watershed imperviousness (Schueler, 1987). In urban streams, higher loadings can cause water quality problems such as turbid water, nutrient enrichment, bacterial contamination, organic matter loads, toxic compounds, temperature increases, and increased instances of trash or debris.

Changes in Stream Habitat and Ecology. The ecology of urban streams is shaped and molded by extreme shifts in hydrology, geomorphology, and water quality that accompany the development process. Stresses on the aquatic community of urban streams are often manifested as

- A shift from external (leaf matter) to internal (algal organic matter) stream production;
- A reduction in diversity in the stream community; and
- A destruction of freshwater wetlands, riparian buffers, and springs.

Structure of These Changes. Stream hydrology, stream form, water quality, and stream habitat and ecology, in order of importance, provide a structure for a designer to develop an integrated picture of the effects of hydrologic alteration on receiving environments (Snodgrass *et al.,* 1996). This structure can also be used to assist in defining environmental and BMP monitoring (Chapter 3) and stream protection (Schueler, 1992) strategies.

EFFECTS OF URBANIZATION ON OTHER RECEIVING ENVIRONMENTS. Effects of Urbanization on Wetland Systems. Neglecting direct intrusion, sound, and other effects outside the sphere of urban hydrology, the effects of urbanization on wetlands can be broadly categorized into physical/geochemical and biological effects. Primary physical/geochemical effects include

- Changes in hydrology and hydrogeology;
- Increased nutrient and other contaminant loads;
- Changes in atmospheric inputs (through increased air emissions to the urban airshed); and
- Compaction and destruction of wetland soil.

Primary biological effects associated with urban stormwater discharges are

- Changes in wetland vegetation;
- Changes in or loss of habitat;
- Changes in the community of organisms (diversity, kind, density);
- Loss of particular biota; and
- Permanent loss of wetlands.

In addition, socioeconomic changes may result, such as decreased passive recreational use. However, through the proper application of BMPs, much of this effect can be mitigated and a healthy ecosystem can be promoted.

Effects of Stormwater Runoff on Lakes. Compared to streams, lakes have at least three inherent differences that cause different reactions to stormwater runoff. These differences are as follows:

- Lakes, because of their volume, respond primarily to the mass of constituents and volume of flow rather than constituent concentration and peak flow rate. The response time is on the order of days or weeks, whereas a stream responds within hours or days.
- After visible refuse and damage, nutrient enrichment and the resulting increase in primary productivity is the most visible sign of urbanization.
- Lakes do not flush as quickly as streams and are net-depositional environments. As such, they act as sinks for sedimented materials and take longer to recover from contamination than do streams.

Heavy metals that adsorb onto sediment particles in urban runoff may not pose an immediate threat to lake ecosystems if the bottom is aerobic. But, if the bottom eventually becomes anaerobic, or is seasonally anaerobic, metals may solubilize in sufficient quantity to be toxic to benthos or pelagic species that swim through the area.

A common effect of urban runoff on lake ecosystems is that the sediment load in the inflowing stream(s) drops out near the inlet and impacts the biota living on the bottom. Depending on the areal extent of the deposition, a sufficient percentage of the natural benthos might be destroyed to alter the food chain, hence the lake organism assemblage.

Another common effect on lakes is that floatables carried into the lake by the runoff stream are blown onto the shore or into small pocket embayments, impairing the aesthetic value of the water body.

A third common effect is increased algae production by the lake if a formerly forested watershed is urbanized. Eventually, summer populations get sufficiently high to create blooms that are aesthetically displeasing. At this point, a lake's value as a drinking water source is also compromised. Blooms may deplete lake oxygen supplies sufficiently to cause fish kills, but ecologically the most significant effect is that the increased primary productivity results in an aquatic environment with decreased diversity and increased "trash" fish populations. Once this occurs, a simple stopping of the source of nutrients will have a limited short-term effect on biota—a polluted lake may take decades to recover naturally.

The broader effects of urbanization on lakes include inputs from storm sewers, CSOs (where they exist), and wastewater treatment plants. Perceptions about increased algal production after urbanization are caused more often by WWTP discharges than by storm sewer discharges. Well-developed loading estimates from the different sources on a watershed basis are needed to adequately distinguish the relative importance of the different influences of urbanization.

Effects of Urbanization on Groundwater. Urbanization affects both groundwater flow and groundwater quality, although, as noted previously, the effects on flow do not always follow expected norms. The primary effects of urbanization on groundwater quality are caused by leaking or leaching of toxic or hazardous substances from significant industrial operations (such as landfills or specific manufacturing sites), gasoline stations, and leaching of previously contaminated soils by infiltrating rainwater. These sources are often the main focuses for soil quality and groundwater quality management in urban areas but are beyond the scope of this manual.

A principal focus of this manual is the way groundwater quality is affected by infiltrating stormwater runoff and BMP facility water. The effects of these sources can be subdivided into perceptions and documented data. The perception of many individuals working in urban stormwater management is that the constituent concentration levels in stormwater runoff (see Tables 1.4 and 1.5) should affect groundwater by increasing the concentration of constituents such as heavy metals, pesticides, and herbicides. Limited data for heavy metals (such as copper, iron, and zinc) suggest that they are sorbed by soils in the bottom of BMP facilities, provided the sediments remain aerobic and have high redox potential (Yousef *et al.*, 1985) and, therefore, do not migrate in a significant way. Hydrophobic organic compounds, such as polychlorinated biphenyls and PAHs, typically do not migrate because of their highly sorptive properties.

However, highly soluble substances such as chlorides and nitrates will move with infiltrating water and not be sorbed. Some evidence is evolving to suggest that groundwater below cities such as Toronto, which uses deicing chemicals, may attain chloride levels between 500 and 800 mg/L (Howard *et al.*, 1991).

Effects of Stormwater Runoff on Estuaries. Because of the relative size of receiving environments, attributing effects to stormwater is more likely in cases of discharges to creeks (rivers) than in discharges to estuaries. To date, few studies have focused solely on the effects of urban runoff to estuaries (Odum and Hawley, 1986, and Jones, 1986). One reason for this is that, in large estuaries, urban runoff effects are inexorably and synergistically associated with the effects of dredging, chemical spills, point source contaminants, and other serious alterations. Therefore, separating the effect of urban runoff is difficult and perhaps artificial.

The effect of urban runoff is also discontinuous. Because episodes of heavy runoff with high pollutant levels are interspersed with long periods of little or no runoff, erosion of near-shore sediments during runoff events is followed by tranquil periods when biological communities can reestablish themselves. In addition, tidal processes spread discharges in multiple directions.

However, there are some definable and measurable effects of urbanization on estuaries. These include

- Sedimentation in estuarial streams;
- Changed hydroperiod of saltwater wetlands, which results from larger, more frequent pulses of fresh water and longer exposure to saline waters because of reductions in base flow; and
- Short-term salinity swings in small estuaries caused by the increased volume of runoff; this affects the local ecosystem, which may be a "delicate" or key reproduction area.

Research studies, hypothesis of effect (Hunsaker and Carpenter [Eds.], 1990), and careful segregation of scale may provide the database necessary for systematically documenting effects. Alternatively, generalized land-use relationships between the lands of urban areas and adjoining estuaries in specific ecozones may be the scale at which effects are documentable. At such a scale, the cumulative effects of urban land development, including stormwater discharges, are related to receiving water characteristics, although the specific effects of stormwater discharges are not measured.

EVOLVING ISSUES

ECOSYSTEMS AND WATERSHED BASIS FOR MANAGEMENT. Water quality management has historically focused on the control of specific pollutants. Control of BOD discharged to streams and estuaries to maintain satisfactory oxygen levels has been evaluated using assimilative capacity models (Streeter and Phelps, 1925). Discharges of phosphorus to lakes are managed to maintain a specific trophic status for lakes (Reckhow and Chapra, 1983). However, some constituents discharged to a receiving water cannot be assimilated and are exported to adjacent ecosystems through cycles of water and elements.

Ecological limits posed by the biosphere and the pollution of specific aquatic systems are leading to attempts to use ecological targets (endpoints) as approaches for defining objectives for urban stormwater quality management. In simple terms, the biosphere is the life-supporting system on the surface of the earth that provides energy, fresh air, potable water, uncontaminated food, and the recycling of wastes. Water and chemical elements cycle through various biological niches and hydrological and geological reservoirs in the biosphere. However, urbanized human populations exert severe stress on specific local biota and natural resources. In addition, the elemental and hydrological cycles cause stress to be exported to adjacent lakes, streams, estuaries, and ecosystems.

Therefore, the ecosystem has become the focus for management of the environment. This requires an understanding of what is meant by the term "ecosystem" and an approach for using the ecosystem as the basis for urban stormwater quality management.

These broad concepts are applied to urban stormwater management issues in the following sections, which consider the overall effects of urbanization on the environment, watersheds, application of ecosystem principles to BMP planning, and impact assessment.

EVOLVING APPROACHES TO STORMWATER QUALITY MANAGEMENT. Urban stormwater quality management is subject to analysis through several different approaches, namely, environmental resource protection planning, comprehensive urban runoff quality management, and comprehensive stormwater management. As it applies in this manual, the term "urban" includes developed and urbanizing areas. In broader terms, evolving approaches to stormwater quality management in North America include

- Water-quality-based mandates for stormwater releases (that is, stormwater effluent quality levels developed by an impact assessment methodology or for a design storm by the stipulation of a defined concentration target for stormwater releases);
- Best management practice plans for stormwater runoff from a specific facility or subwatershed;
- A comprehensive stormwater quality protection strategy that includes BMP plans;
- A comprehensive receiving water protection strategy; and
- Watershed-based approaches.

The comprehensive stormwater quality protection strategy includes pollution prevention (also called source control), pollution control, land-use planning, and regulatory control.

Pollution control, in the form of treatment-type BMPs, has been a primary focus of recent efforts. Because most BMPs are passive in nature, protection strategies have expanded to include pollution prevention and regulatory control approaches. However, the effectiveness of pollution prevention efforts are often not known quantitatively. With the development of data that link percent imperviousness thresholds to stream degradation in headwater areas (see Figures 1.2 and 1.3), the potential for using land-use planning as a tool has commonly been proposed.

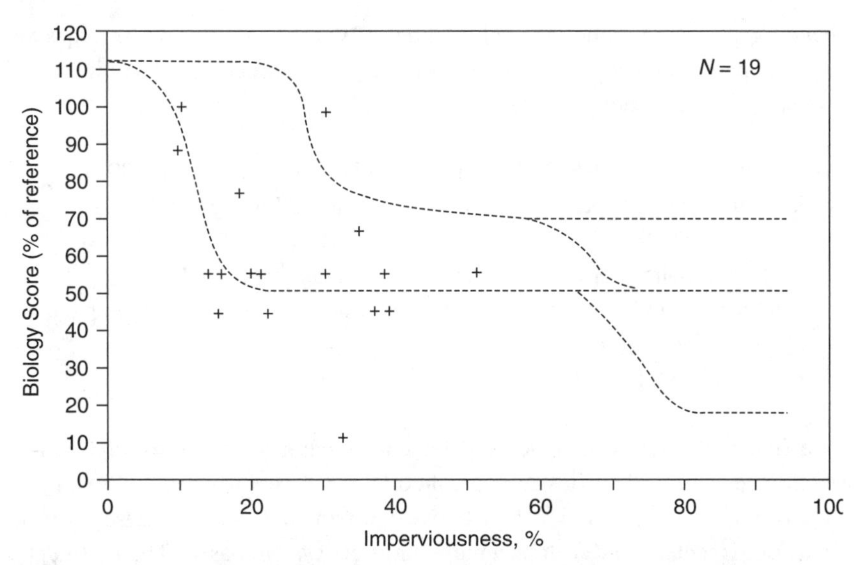

Figure 1.3 Effect of watershed imperviousness on biologic integrity within the Northern Piedmont ecoregion of Delaware, 1993.

Both pollution prevention and pollution control are cornerstone approaches for stormwater quality management. They are addressed in this manual in Chapters 4 and 5, respectively. Watershed approaches historically have been used for flood control and erosion control. Future direction will come from ecosystem analysis on a watershed basis and attempts to integrate these concepts with flood control, erosion control, and water quality management.

THE WATERSHED AS NATURAL BIOPHYSICAL BOUNDARY. An ecosystem approach to planning requires that boundaries for examining relationships between the natural environment and human activities be based on biophysical rather than political issues. To this end, watersheds and subwatersheds provide a biophysical basis for planning and management.

A watershed uses the hydrologic cycle as the pathway that integrates the physical, chemical, and biological processes of the basin ecosystem. It provides a fundamental unit with real boundaries of ecological significance. This unit provides a quantitative reference frame to make inferences on ecological stress, instream flow, and the cumulative effects of land-use development. A watershed basis has been the natural approach for water-quantity-related issues (such as flood control or erosion control) and, more recently, for water quality issues (for example, dissolved oxygen control in streams and phosphorus loadings to streams).

The ecosystem approach assists in balancing economic, social, and environmental needs. The approach purposefully links aquatic, terrestrial, and human life within the watershed as a healthy self-sustaining ecosystem. The benefits of this are numerous and long term. Watershed planning becomes an integral part of an ecosystem and land-use planning process, set within the context of environmental sustainability.

WATERSHED PLANNING. Watershed planning (U.S. EPA, 1993), described further in Chapter 2, involves at least the following three phases:

- Background review and definition of management goals;
- Analysis and development of a plan (strategies including an impact assessment); and
- Implementation of a plan.

Watershed planning is an essential element for environmental resource protection because it identifies ecological components requiring protection, systematically analyzes relationships between urban land uses and these ecological components, and develops a plan that is ecosystem based. The pertinent steps for watershed planning are outlined in Table 1.7; methods for developing an overall watershed approach are described in Chapter 2.

Representative products (or deliverables) expected from a watershed study, a subwatershed study, and a BMP facilities study are given in Table 1.8. An ecosystem approach requires that a master environmental plan be developed in addition to "and integrated with" a master drainage plan. A fundamental problem, however, is how to implement an ecosystem approach to planning on a watershed basis. An ecosystem approach forces a designer to look not only at air, water, land, and living things but also at the interrelationships of these components. Furthermore, the study of these interrelationships includes both "quality" and "quantity" aspects.

Traditionally, science has taken a reductionist approach by studying the individual components of a system. In the context of a watershed, this would mean studying streams, groundwater, aquatic communities, and other environmental components of interest. It also includes studying municipal WWTP discharge, industrial waste sources, CSOs, sanitary sewer overflows, and stormwater discharges. However, ecological planning requires an understanding of how these components interact. This study of interrelationships and the whole is known as "integration."

USE OF AN ECOSYSTEM APPROACH FOR PLANNING. Best management practice planning occurs on both a watershed scale and a site scale (see Table 1.9). Watershed-level plans are more suited to incorporating ecosystem-based geographic boundaries, multiple project influences and effects, and the setting of ecosystem health objectives with the assistance of public input.

However, application of an ecosystem approach for individual development projects and BMP facilities is more difficult because it is harder to understand the relationship of a specific project to the surrounding ecosystem

Table 1.7 Steps in watershed planning.

Phase I background information
Define study goals
Inventory watershed and establish VECs/human uses
Design/initiate necessary field work
Public participation
Define present land uses and pollution sources
Define present state of the environment
Define future causes of environmental degradation (land-use changes, population growth, emission sources)
Define management goals for aquatic systems/human uses
Phase II analysis and development of plan
Define ecosystem framework
Define impact hypothesis (modes of impact, exposure pathways, environmental changes)
Define ecological/environmental endpoints and targets (evaluation criteria)
Define by screening studies, magnitude of impact
Define potential ecosystem protection/mitigating measures (such as BMPs, conservation practices)
Prioritize analysis
Define and apply prediction techniques
Establish alternative management strategies
Review effectiveness, uncertainties, cost, benefits, risks, implementation
Assess whether management goals/environmental targets can be met
Public and institutional review
Select and develop recommended watershed plan
Define uncertainties
Financing
Phase III implementation plan
Define implementation strategy
Define uncertainties and necessary research
Define lead agency for implementing different components of plan
Define funding sources
Define approval mechanisms
Define direction for subwatershed and BMP plans
Define schedule and phasing
Establish implementation committee/task force and terms of reference
Define long-term monitoring program

Table 1.8 Outline of deliverables expected from watershed planning.

Deliverables expected from watershed study
- Inventory of resources and establishment of management objectives
- Statement of environment endpoints and quality targets
- List of BMPs and mitigating measures and location in watershed
- List of impact statements and uncertainties to be addressed in future monitoring programs

Deliverables expected from subwatershed study
- Detailed evaluation of watershed information
- More detailed analysis of environmental quality targets (defined in watershed study) and land-use densities and mitigating measures required to achieve the targets
- Detailed hydrology and hydraulic modeling (as required) to define sizes of BMP facilities required for hydrological aspects
- BMP selection

Deliverables expected from BMP facilities study
- Confirmation of critical hydrological/hydraulic assumptions used previously
- Selection of ecosystem features at BMP facility
- Conceptual followed by detailed layout of BMP/BMPs for upstream, site, and downstream aspects
- Detailed design

and it is difficult to assess the cumulative effects that are involved. Therefore, for individual projects in which watershed plans have not been developed, an ecosystem approach should include

- Establishing ecological objectives based on desired beneficial uses;
- Considering effects beyond the vicinity of the facility site but within the watershed (another definition is to "look upstream," "downstream," and "at the point of release" of the stormwater);

Table 1.9 Proposed hierarchy based on geographic/geomorphic scaling considerations.

System level	Linear spatial scale, m	Aerial spatial scale, m^2	Time scale of continuous potential persistence, years	Time scale of persistence under human disturbance patterns, years
Watershed	10^5	10^{10}	10^6–10^5	10^4–10^3
Subwatershed	10^4	10^8	10^4–10^3	10^2–10^1
Reach	10^2–10^1	10^5	10^2–10^1	10^1–10^0
Site	10^1–10^0	10^2	10^0	10^0–10^{-1}
Habitat element	10^0–10^{-1}	10^1	10^0–10^{-1}	10^{-1}–10^{-2}

- Establishing the boundaries of the study to include off-site areas with functional linkages on site, sometimes by facilitating cooperation among landowners to include stream corridors, wildlife management areas, and local groundwater aquifers;
- Increasing collaboration among the team members dealing with biology, surface water, and hydrogeology;
- Illustrating ecosystem functions and linkages in "pathway diagrams" for project effects such as air, noise, litter, and water quality and using these to scope the assessments;
- Systematically considering interactions among components of the environment (for example, the effects of surface water and groundwater on biology);
- Describing existing and past conditions, including both the functions of the ecosystem components and their interactions with one another (for example, past construction of a storm sewer has dried up a wetland or the construction of barriers has reduced stream flow); and
- Systematically considering environmental effects of other existing and future facilities or projects in assessing the net effects of the undertaking.

The first four points are often considered in watershed evaluations; the other points represent the evolving practice toward an "ecosystem" approach.

ENVIRONMENTAL MONITORING AND IMPACT ASSESSMENT

Methods for conducting an impact assessment include evolving methodologies such as a stress-response framework (Hunsaker and Carpenter [Eds.], 1990), risk assessment approaches (Suter, 1993, and Warren-Hicks *et al.*, 1988), environmental effects monitoring (Hollings [Ed.], 1978, and Hunsaker and Carpenter [Eds.], 1990) and other methodologies detailed in Chapter 3.

Furthermore, watershed studies, research studies, and experience now permit desired environmental quality goals to be stated in physical, biological, and chemical terms. However, cause–effect methodologies for relating BMP performance to biotic changes are only now evolving. Therefore, impact assessment methodologies presented in Chapter 3 are stated in terms of hydrological performance (using hydrological models); loadings of constituents to receiving environments; and bioassessment, biocriteria, and inferences that can be gathered from field data rather than from deterministic cause–effect relationships.

EVOLVING APPROACHES TO BIOLOGICAL IMPACT ASSESSMENT. Environmental Endpoints. The term "environmental endpoint" has evolved from the field of ecological risk assessment (U.S. EPA, 1992, and Warren-Hicks *et al.*, 1988) to include both "ecological entities" and human uses/concerns (such as human health and safety and property protection). An environmental endpoint can be defined as

- A component of the environment (for example, air or soil quality) that is valued by the public and represents a significant attribute of the environment;
- A part of the environment that can be measured (often using various biological, chemical, or physical parameters) or calculated/modeled;
- An entity that is affected by urbanization or a specific proposed project; or
- An intermediate or final compartment along a series of environmental pathways that are affected by the project.

The following environmental endpoints and methods are presently being applied in the field of impact assessment:

- *In situ* properties of receiving environments:
  - Hydrological characteristics (such as flow and velocity);
  - Physical and chemical constituents of surface water;
  - Physical and chemical characteristics of sediments (Livingston *et al.,* 1995);
  - Water column communities (such as algae, zooplankton, and fish);
  - Benthic communities;
  - Human health characteristics (such as fecal coliforms and viruses);
  - Other biological and toxicity characteristics (for example, BOD); and
  - Morphometric properties (particularly in streams).
- Laboratory-based measurements:
  - Aquatic toxicity,
  - Genotoxicity,
  - Sediment toxicity, and
  - Chemical concentrations of field samples.
- *In situ* probes and biomonitors:
  - Algal organisms to measure bioaccumulation or *in situ* toxicity,
  - Attached algae that sorb particular contaminants, and
  - Young-of-year fish that migrate limited distances in stream.
- Stormwater and BMP discharge:
  - Chemical constituents, and
  - Biological properties (BOD, fecal coliforms, toxicity).

The category "*in situ* probes" involves putting organisms in place for brief periods of time to measure phenomena not measured by typical chemical procedures or using special *in situ* organisms to detect effects.

Biologically Based Effects Monitoring. These types of environmental endpoints represent a shift away from a chemical concentration approach for measuring environmental quality and toward a biologically based effects approach (for example, bioassessment or biocriteria). This change is occurring because biological measurements provide more meaningful data for characterizing environmental quality. However, data obtained are often difficult to interpret. Moreover, while water quality standards for many parameters provide a reference by which to compare water characteristics, water quality objectives for many biomonitors are not known. By installing biomonitors above and below stormwater discharges in an urban area, an increase in body burden for a particular chemical parameter can be measured, although its environmental meaning is unclear.

An environmental effects monitoring program can be thought of as having the following structure in its biological and chemical components:

- Concentrations of constituents and whole water toxicity of the BMP-discharged water;
- Sediment accumulation and toxicity;
- Bioaccumulation (for example, contaminant burden in organism tissue or human consumptive concerns); and
- Trophic-level effects (such as subcellular, cellular, community level, and population level).

The development of monitoring systems during the next decade may be aimed at formulating a balance between chemical- and biological-monitoring approaches in this evolving science. A set of principles and approaches for monitoring the effectiveness of stormwater quality BMP and receiving water effects is provided in Chapter 3.

Focus in This Manual. In the evolving field of environmental monitoring, the following physical and chemical factors are particularly relevant monitoring components (Urbonas, 1994) as they apply to information presented in this manual:

- Hydrological alterations/mitigations to stormwater runoff from an urban catchment provided by the BMP;
- The treatment efficiency provided by the BMP for SS and attached contaminants;
- Nutrient removal provided by the facility; and
- Temperature properties of discharged waters.

These monitoring components are presently carried out at many sites and are discussed in detail in Chapter 3 in relation to monitoring BMPs and their receiving waters.

In addition to physical and chemical factors, important environmental monitoring factors for biological components include

- Toxicological properties of runoff water (as a laboratory procedure) and scale-up to receiving water consequences;
- Bioaccumulation of contaminants in the food chain; and
- Alteration of the benthic community because of the physical stress of runoff water.

A key technical issue of monitoring is to determine the relative importance of flow alterations (physical, morphological changes that degrade the benthic community) and chemical effects (through toxicity and bioaccumulation) on determining the primary cause of effects on receiving water ecosystems. Presently, emphasis is placed on the structure and abundance of water column and benthic communities in receiving environments rather than on cause–effect relationships. If either the physical or chemical effects of urban stormwater runoff were clearly shown to be the primary causes of degradation of urban receiving water ecosystems, improvement in the rationale for stormwater BMP design could be achieved.

ISSUES OF SCALE. This section examines the influence of scale on the response time of streams to urbanization and the receiving water scale on which monitoring can be expected to detect changes because of the installation of one BMP facility or a set of facilities within a watershed.

Physical and Temporal Response of Streams. The geomorphological alteration of streams caused by altered urban hydrology has one of the longest response times of all receiving water attributes. In terms of the scale of a stream reach, at least a decade may be required for a stream ecosystem to respond to altered hydrology.

Scale for Monitoring the Response of Streams. Research has suggested that a threshold effect is observed in the relationship between imperviousness of a watershed and a stream's morphology and ecosystem. The relationship between imperviousness and benthic organism density, presented in Figure 1.3, shows a significant change in watersheds having 10 to 20% imperviousness. Shaver *et al.* (1995) hypothesized that installations of BMPs might protect the stream and change the threshold to 30 to 50% (see Figure 1.3).

Because the streams shown in Figure 1.3 are largely unprotected with BMPs, validation of this hypothesis should be sought by comparing subwatersheds fully protected by BMPs to equivalent watersheds without BMPs.

In developing such a monitoring program, additional points should be considered. Benthic organisms, if used as the response variables, integrate the effects of hydrologic control and water quality protection provided by BMPs. As discussed above, a monitoring program that seeks to separate the effects of water quality improvement from those of hydrologic control requires a special design. Secondly, many of the streams represented in research drain small watersheds (typically containing first- and second-order streams). Therefore, different threshold effects may be measured in larger streams.

TOTAL EFFECTS METHODOLOGIES. Broader based planning methodologies and environmental assessment are evolving to account for the total impacts of urbanization on the environment. Best management practice impact assessment methodologies may need to consider such new issues especially in multifaceted watershed studies.

Other Ecosystem Boundaries. The watershed boundary is an appropriate ecological boundary for managing stormwater runoff from catchments discharged to receiving environments (surface waters and their contained ecosystems, and groundwater). But, other boundaries may need to be considered.

The watershed boundary is not the only boundary used in ecosystem analysis and planning. In fact, there is no single, all-inclusive ecosystem approach to analysis and planning. Rather, there are several approaches in existence, such as landform based (such as the Appalachian mountains), ecological land classification based (ecoregions and ecozones) (see Livingston *et al.*, 1995), landscape and natural heritage, groundwatersheds, and airsheds. Human concerns include urban regions and commutersheds. An important points is that we are the ones identifying the boundaries of ecosystems as a geographical basis for management. However, ecosystems function regardless of our efforts to compartmentalize and spatially define them. The ecosystems that we map are simply "constructs" to facilitate our understanding and management. The chosen boundaries are seldom "absolute," but, rather, should be regarded as "diffuse" and as an approximation of what is appropriate for a given concern, theme, and chosen scale.

Total Impacts of Urbanization. This chapter has concentrated on the environmental impacts of urbanization caused by stormwater runoff and the mitigation provided by urban BMPs. However, stormwater runoff does not account for all the environmental impacts of urbanization. This section outlines the broader issues to assist the designer in understanding other evolving issues that may influence their decisions related to stormwater management. The broader issues include direct spatial effects involving loss of lands; indirect impacts through pathways, especially hydrological and ecological alterations; and cumulative effects.

DIRECT SPATIAL EFFECTS. Loss of lands and direct alteration of stream channels (direct spatial effects) have historically been among the largest effects of urbanization on surface waters and their contained ecosystems. In fact, a permanent feature of urban development is the loss of natural features, valued land forms, groundwater recharge/discharge areas, and wildlife habitats if development occurs on lands occupied by these land uses. In some literature, these effects are called "form effects" or "spatial impacts," in the geographical information system sense of "spatial analysis."

Spatial impacts resulting from urbanization include reconstruction, encroachment, and intrusion by humans. For receiving environments such as stream corridors and valleys, direct effects have historically been caused by enclosing streams in pipes or open channels, channel realignments, and stream crossings by infrastructure. Stormwater impacts are not easily addressed under this topic because they are an indirect effect.

PATHWAYS. Pathway-based methodologies are used in fields such as ecological and human health risk assessment. Pathways involve the movement of water or contaminants from industrial, residential, commercial, or agricultural areas through atmospheric, surface water, or groundwater routes (indirect impacts through pathways) to affect biota in the various ecosystems where they live, breed, and raise their offspring. These ecosystems units are called habitats. The pathways may be physical or biological in nature. Pathways relevant to urbanization issues include hydrological pathways, biological pathways, and air pathways.

Stormwater runoff is one pathway used to analyze the impacts of urbanization and is considered in this manual. As summarized above, the main effects of urbanization on hydrological pathways include increases in surface water flow, reduction in infiltration to groundwater, increased risk of water quality degradation, increases in flooding and erosion potential, and habitat degradation and even destruction.

The main effects of biological pathways include the bioaccumulation of toxic elements through the food chain. The main effects on air pathways include increased emissions of various contaminants to air from fuel consumption for heating residences, industrial and commerical buildings, transportation, and industrial operations. These latter pathways are generally not considered in this manual.

CUMULATIVE EFFECTS. Cumulative effects assessment is an additional factor required in recent planning and monitoring studies. The current system for reviewing planning applications, particularly in rural areas within the urbanizing fringe, is primarily oriented to site-specific analysis and, therefore, does not anticipate the broader, longer term environmental implications of permitting many individual sites to be developed. In some planning circles, these are known as cumulative environmental effects (for additional discussion, see Constant and Wiggins, 1991).

Watershed studies and ecosystem-based planning are examples of cumulative effects analyses. They allow one to examine the effect of all development on loadings of contaminants to surface waters and groundwater; changes in groundwater and surface water flow; changes in an instream or lacustrine fishery habitat caused by changes in surface water flow or water quality; and changes in other ecosystems or their habitat because of the introduction of contaminants through air or water pathways.

However, few studies have been carried out to date to measure (monitor) the cumulative effects of implementing urban BMPs throughout a watershed, a deficiency that needs to be addressed in future research.

REFERENCES

American Society of Civil Engineers and Water Environment Federation (1992) *Design and Construction of Urban Stormwater Management Systems.* Am. Soc. Civ. Eng. Manual and Report of Engineering Practice No. 77, New York, N.Y.; Water Environ. Fed.Manual of Practice No. FD-20, Alexandria, Va.

Bannerman, R., *et al.* (1984) *Evaluation of Urban Nonpoint Source Pollution Management in Milwaukee County, Wisconsin.* U.S. EPA, Region V, Chicago, Ill.

Barrett, M.E., *et al.* (1995) Water Quality Impacts of Highway Construction and Operation in Texas. *TR News*, **179,** 15.

Belore, H. (1991) Regionalization of Low Flows in Ontario. *Proc. Environ. Res. Technol. Trans. Conf.*, Ont. Ministry Environ., Toronto, Ont., Can.

Bengtsson, L. (1982) Snowmelt-Generated Runoff in Urban Areas. In *Urban Stormwater Hydraulics and Hydrology.* B.C. Yen (Ed.), Water Resources Publications, Littleton, Colo.

Booth, D. (1991) Urbanization and the Natural Drainage System—Impacts, Solutions and Prognoses. *Northwest Environ. J.,* **7**, **1,** 93.

Booth, D., and Reinelt, L. (1993) Consequences of Urbanization on Aquatic Systems—Measured Effects, Degradation Thresholds, and Corrective Strategies. *Proc. Watershed Manage.,* Alexandria, Va.

Carter, R.W. (1961) Magnitude and Frequency of Floods in Suburban Areas. U.S. Geol. Surv., Washington, D.C.

Cheng, G., *et al.* (1994) Proposed Monitoring Program Design for City of Austin. Poster presented at NPDES Related Monit. Needs Conf. Eng. Foundation, Crested Butte, Colo.

City of Olympia (1994b) Impervious Surface Reduction Study. Public Works Dep., Olympia, Washington.

Constant C.K., and Wiggins, L.L. (1991) Defining and Analyzing Cumulative Environmental Impacts. *Environ. Impact Assessment Rev.*, **11,** 297.

Cooke, T., *et al.* (1995) Stormwater NPDES Monitoring in Santa Clara Valley. In *Stormwater NPDES Related Monitoring Needs.* H.C. Torno *et al.*

Cowherd, C., Jr., *et al.* (1977) *Quantification of Dust Entrainment from Paved Roadways*. EPA-450/3-77-027, U.S. EPA, Research Triangle Park, N.C.

Craig, G.J., and Tracy, H. (1993) Ultraviolet Disinfection of Stormwater at Longfields/Davidson Heights Stormwater Treatment Facility. In *Stormwater Management and Combined Sewer Control Technology Transfer Conference.* Environ. Can., Wastewater Technol. Cent., Burlington, Ont., Can., 129.

Department of City Development (1981) Fugitive Dust Emissions: Their Sources and Their Control in Milwaukee's Menomonee River Valley. Milwaukee, Wis.

Driscoll, E.D., and James, W. (1987) Evaluation of Alternate Distributions. In *Pollution Control Planning in Ontario.* W. James (Ed.), Computational Hydraulics, Inc., Guelph, Ont., Can., 139.

Driscoll, E.D., *et al.* (1990) *Pollutant Loadings and Impacts from Highway Stormwater Runoff.* Vol. I and Vol. III, Office Eng. and Highway Oper. Res. Dev., Fed. Highway Admin., McLean, Va.

D'Andrea, M.D., *et al.* (1993) Characterization of Stormwater and Combined Sewer Overflows in Metropolitan Toronto. *Proc. Stormwater Manage. Combined Sewer Control Technol. Transfer Conf.,* Wastewater Technol. Center, Burlington, Can.

Fam, S., *et al.* (1987) Hydrocarbons in Urban Runoff. *J. Environ. Eng.,* **113,** 5, 1032.

Field, R., *et al.* (1974) Water Pollution and Associated Effects from Street Salting. *J. Environ. Eng.,* **100,** EE2, 459.

Förster, J. (1990) Roof Runoff: A Source of Pollutants in Urban Storm Drainage Systems. *Proc. 5th Int. Conf. Urban Storm Drainage.* Y. Iwasa and T. Sueishi (Eds.), Osaka Univ., Jpn., 469.

Glenne, B. (1984) Simulation of Water Pollution Generation and Abatement on Suburban Watershed. *Water Resour. Bull.,* **20,** 2, 211.

Halverson, H.G., *et al.* (1982) Runoff Contaminants from Natural and Manmade Surfaces in a Nonindustrial Urban Area. *Proc. Int. Symp. Urban Hydrol., Hydraul. Sediment Control,* Univ. Ky., Lexington, 233.

Halverson, H.G., *et al.* (1984) Contribution of Precipitation to Quality of Urban Storm Runoff. *Water Resour. Bull.,* **20,** 6, 859.

Harremöes, P. (1988) Stochastic Models for Estimation of Extreme Pollution from Urban Runoff. *Water Res.* (G.B.), **22,** 1017.

Harrison, R.M., and Johnston, W.R. (1985) Deposition Fluxes of Lead, Cadmium, Copper and Polycyclic Aromatic Hydrocarbons on the Verge of a Major Highway. *Sci. Total Environ.,* **46,** 121.

Heaney, J.P., and Huber, W.C. (1973) Stormwater Management Model, Refinements, Testing and Decision-Making. Dep. Environ. Eng. Sci., Univ. Fla., Gainesville.

Herricks, E.E. (1991) Impacts of Urban Stormwater Runoff on Receiving Waters. Paper presented at Conf. Stormwater Manage., Computational Hydraulics, Inc., Guelph, Ont., Canada.

Herricks, E.E., *et al.* (1994) Time-Scale Toxic Effects in Aquatic Ecosystems. In *Stormwater NPDES Related Monitoring Needs*. H.C. Torno *et al.* (Eds.), Eng. Foundation, Am. Soc. Civ. Eng., New York N.Y., 353.

Hoffman, E.J. (1985) Urban Runoff Pollutant Inputs to Narragansett Bay: Comparison to Point Sources. In *Proc. Natl. Conf. Perspectives Nonpoint Pollut.* EPA-440/5-85-001, U.S. EPA, Washington, D.C.

Hoffman, E.J., *et al.* (1984) Urban Runoff as a Source of Polycyclic Aromatics to Coastal Waters. *Environ. Sci. Technol.,* **18,** 580.

Hollings, C.S. (Ed.) (1978). *Adaptative Environmental Assessment and Management.* John Wiley and Sons, Inc., New York, N.Y.

Hollis, G. (1975) The Effect of Urbanization on Floods of Different Recurrence Intervals. *Water Resour. Res.,* **11,** 3, 431.

Howard, K.H., *et al.* (1991) Groundwater Contamination by Road Deicing Salts— Implications on a Salt Balance Performed on Highland Creek, a Creek in Metropolitan Toronto. *Proc. Environ. Res. Technol. Trans. Conf.*, Ont. Ministry Environ., Toronto, Ont., Can.

Huber, W.C. (1986) Deterministic Modeling of Urban Runoff Quality. In *Urban Runoff Pollution.* H.C. Torno *et al.* (Eds.), Springer Verlag, Berlin, Ger., 167.

Hunsaker, C.T., and Carpenter, D.E. (Eds.) (1990) *Ecological Indicators for the Environmental Monitoring and Assessment Program.* EPA-600/3-90-060, U.S. EPA, Office Res. Dev., Research Triangle Park, N.C.

Hydrologic Engineering Center (1975) Urban Stormwater Runoff—STORM. U.S. Army Corps Eng., Davis, Calif.

James, W., and Boregowda, S. (1986) Continuous Mass Balance of Pollutant Build-Up Processes. In *Urban Runoff Pollution.* H. Torno *et al.* (Eds.), Springer Verlag, Berlin, Ger., 243.

Jones, J.E. (1986) Urban Runoff Impacts on Receiving Waters. In *Urban Runoff Quality—Impact and Quality Enhancement Technology.* B. Urbonas and L. Roesner (Eds.), Am. Soc. Civ. Eng., New York, N.Y.

Kerri, K.D., *et al.* (1985) Forecasting Pollutant Loads from Highway Runoff. Transportation Res. Rec., Transportation Res. Board, Natl. Res. Council, 1017.

Leopold, L.B. (1968) Hydrology for Urban Land Planning—A Guide Book on the Hydrologic Effects of Urban Land Use. U.S. Geol. Surv., Water Supply Paper 1591-C.

Livingston, E.H., *et al.* (1995) Use of Sediment and Biological Monitoring. In *Stormwater NPDES Related Monitoring Needs.* H.C. Torno (Ed.), Eng. Foundation, Am. Soc. Civ. Eng., New York, N.Y., 375.

Lord, B.N. (1988) Program to Reduce Deicing Chemical Use. In *Design of Urban Runoff Quality Controls. Proc. Eng. Found. Conf.,* L.A. Roesner *et al.* (Eds.), Potosi, Mo., 421.

Macrae, C., and Marsalek, J. (1992) The Role of Stormwater in Sustainable Urban Development. *Proc. Can. Hydrol. Symp. Hydrol. Contrib. Sustain. Dev.,* Winnipeg, Man., Can.

Marsalek, J. (1986) Toxic Contaminants in Urban Runoff. In *Urban Runoff Pollution.* H. Torno *et al.* (Eds.), Springer Verlag, Berlin, Ger., 39.

Marsalek, J. (1990) PAH Transport by Urban Runoff from an Industrial City. *Proc. 5th Int. Conf. Urban Storm Drainage.* Y. Iwasa and T. Sueishi (Eds.), Osaka Univ., Jpn., 481.

Marsalek, J., and Torno, H.C. (Eds.) (1993) Urban Storm Drainage. In *Proc. 6th Int. Conf. Urban Storm Drainage Niagara Falls, Ont., Can.,* J. Marsalek and H.C. Torno (Eds.), Seapoint Publishers, Victoria, B.C., Can.

Marsalek, J. (1991) Pollutant Loads in Urban Stormwater: Review of Methods for Planning-Level Estimates. *Water Resour. Bull.,* **27,** 283.

Marsalek, J., *et al.* (1993) Urban Drainage Systems: Design and Operation. *Water Sci. Technol.,* **27,** 12, 31.

McComas, M.R., *et al.* (1976) A Comparison of Phosphorus and Water Contributions by Snowfall and Rain in Northern Ohio. *Water Resour. Bull.,* **12,** 3, 519.

Ng, H.Y.F. (1987) Rainwater Contribution to the Dissolved Chemistry of Storm Runoff. In *Urban Storm Water Quality, Planning and Management.* W. Gujer and V. Krejci (Eds.), IAHR-IAWPRC, École Polytechnique Fédérale, Lausanne, Switz., 21.

Novotny, V. (1987) Effect of Pollutants from Snow and Ice on Quality of Water from Urban Drainage Basins. Dep. Civ. Eng., Marquette Univ., Milwaukee, Wis.

Novotny, V. (1988) Modeling Urban Runoff Pollution During Winter and Off-Winter Periods. In *Advances in Environmental Modelling.* A. Marani (Ed.), Elsevier, Amsterdam, Neth., 43.

Novotny, V., and Chesters, G. (1981) *Handbook of Nonpoint Pollution: Sources and Management.* Van Nostrand Reinhold Company, New York, N.Y.

Novotny, V., and Kincaid, G.W. (1982) Acidity of Urban Precipitation and Its Buffering During Overland Flow. In *Urban Stormwater Quality, Management and Planning.* B.C. Yen (Ed.), Water Resources Publications, Littleton, Colo., 1.

Oberts, G.L. (1986) Pollutants Associated with Sand and Salt Applied to Roads in Minnesota. *Water Resour. Bull.,* **22,** 3, 479.

Oberts, G.L. (1990) Design Considerations for Management of Urban Runoff in Wintry Conditions. *Proc. Int. Conf. Urban Hydrol. Under Wintry Conditions,* Narvik, Nor.

Odum, W.E., and Hawley, M.E. (1986) Impacts of Urban Runoff on Estuarine Systems. In *Urban Runoff Quality—Impact and Quality Enhancement Technology.* B. Urbonas and L. Roesner (Eds.), Am. Soc. Civ. Eng., New York, N.Y.

Olivieri, V.P., *et al.* (1977) *Microorganisms in Urban Stormwater*. EPA-600/2-77-087, U.S. EPA, Munic. Environ. Res. Lab., Cincinnati, Ohio.

Olivieri, V.P., *et al.* (1989) Microbiological Impacts of Storm Sewer Overflows: Some Aspects of the Implication of Microbial Indicators for Receiving Waters. In *Urban Discharges and Receiving Water Quality Impacts.* J.B. Ellis (Ed.), Pergamon Press, Oxford, Eng., 47.

Oswald, W.J., and Golueke, G.G. (1966) Eutrophication Trends in the United States—A Problem. *J. Water Pollut. Control Fed.,* **38**, 964.

Pitt, R. (1979) *Demonstration of Nonpoint Pollution Abatement Through Improved Street Cleaning Practices*. EPA-600/2-79-161, U.S. EPA, Cincinnati, Ohio.

Pitt, R., and Barron, P. (1989) *Assessment of Urban and Industrial Stormwater Runoff Toxicity and the Evaluation/Development of Treatment for Runoff Toxicity Abatement–Phase* I. Rep. for U.S. EPA, Storm Combined Sewer Pollut. Program, Edison, N.J.

Pitt, R., *et al.* (1990a) Analysis of Cross-Connections and Storm Drainage. *Proc. Urban Stormwater Enhancement Source Control, Retrofitting Combined Sewer Technol.,* Am. Soc. Civ. Eng. Eng. Found., Davos, Switz.

Pitt, R., *et al.* (1990b) *Assessment of Non-Stormwater Discharges into Separate Storm Drainage Networks–Phase I: Manual of Practice*. Rep. for U.S. EPA, Storm Combined Sewer Pollut. Program, Edison, N.J.

Randall, C.W., *et al.* (1982) Comparison of Pollutant Mass Loads in Precipitation and Runoff in Urban Area. In *Urban Stormwater Quality, Management and Planning*. B.C. Yen (Ed.), Water Resources Publications, Littleton, Colo., 29.

Reckhow, K.H., and Chapra, S.C. (1983) *Engineering Approaches for Lake Management. Volume 1. Data Analysis and Empirical Modeling.* Butterworth-Heinemann, Stoneham, Mass.

Sartor, J.D., and Boyd, G.B. (1972) *Water Pollution Aspects of Street Surface Contamination*. EPA-R2/72-081, U.S. EPA, Washington, D.C.

Sartor, J.D., *et al.* (1974) Water Pollution Aspects of Street Surface Contamination. *J. Water Pollut. Control Fed.,* **46,** 458.

Schmidt, S.D., and Spencer, D.R. (1986) The Magnitude of Improper Waste Discharges in an Urban Stormwater System. *J. Water Pollut. Control Fed.,* **58,** 744.

Schueler, T.R. (1987) *Controlling Urban Runoff: A Practical Manual for Planning and Designing Urban Best Management Practices*. Metro. Wash. Council Gov., Washington, D.C.

Schueler, T.R. (1992) Mitigating the Impacts of Urbanization. In *Implementation of Water Pollution Control Measures in Ontario*. W.J. Snodgrass and J.C. P'Ng (Eds.), Univ. Toronto Press, Toronto, Ont., Can.

Schueler, T.R. (1994) *Watershed Protection Techniques: A Quarterly Bulletin on Urban Watershed Restoration and Protection Tools.* Center for Watershed Protection, **1,** 1.

Schueler, T.R., *et al.* (1992) *Design of Stormwater Wetland Systems.* Center for Wetland Protection, Silver Spring, Md.

Shaheen, D.G. (1975) *Contributions of Urban Roadway Usage to Water Pollution.* EPA-600/2-75-004, U.S. EPA, Office Res. Dev., Washington, D.C.

Shaver, E., *et al.* (1995) Watershed Protection Using An Integrated Approach. In *Stormwater NPDES Related Monitoring Needs.* H.C. Torno *et al.* (Eds.), Am. Soc. Civ. Eng., New York, N.Y., 435.

Snodgrass, W.J., *et al.* (1996) Can Environmental Impacts of Watershed Scale Development Be Measured? In *Effects of Watershed Development and Management on Aquatic Ecosystems.* L. Roesner (Ed.), Am. Soc. Civ. Eng., New York, N.Y., 351.

Strecker, E.W., *et al.* (1987) Characterization of Pollutant Loadings from Highway Runoff in the USA. In *Urban Storm Water Quality, Planning and Management.* W. Gujer and V. Krejci (Eds.), IAHR-IAWPRC, École Polytechnique Fédérale, Lausanne, Switz., 85.

Strecker, E.W., *et al.* (1990) The U.S. Federal Highway Administration Receiving Water Impact Methodology. *Sci. Total Environ.,* **93,** 489.

Streeter, H.B., and Phelps, E.B. (1925) A Study of Pollution and Natural Purification of the Ohio River. Public Health Bull. 146, U.S. Public Health Serv., Washington, D.C.

Suter, G.W. (1993) *Ecological Risk Assessment.* Lewis Publishers, Boca Raton, Fla.

Thomson, N.R., *et al.* (1995) Prediction and Characterization of Highway Stormwater Quality. R & D Report MAT-94-09, Ont. Ministry Transportation, Toronto, Ont., Can.

Thomson, N.R., *et al.* (1997) Highway Stormwater Runoff Quality: Development of Surrogate Parameter Relationships. *Water, Air, Soil Pollut.*, **94,** 307.

Thon, H.M. (1992) Stormwater Regulations and CSO Strategy in the USA. In *Implementation of Pollution Control Measures for Urban Stormwater Runoff.* W.J. Snodgrass and J.C. P'Ng (Ed.), Univ. Toronto Press, Toronto, Ont., Can.

Torno, H.C., *et al.* (Eds.) (1994) *Stormwater NPDES Related Monitoring Needs.* Am. Soc. Civ. Eng., New York, N.Y.

University of Connecticut (1994) An Overview of Nonpoint Source Pollution Issues. Transportation Inst. Technol. Transfer Center, New Haven, Conn.

Urbanos, B. (1994) Parameters to Report with BMP Monitoring Data. In *Stormwater NPDES Monitoring Needs.* H.C. Torno (Ed.), Am. Soc. Civ. Eng., New York, N.Y., 306.

Urbonas, B.R., and Roesner, L.A. (1992) Hydrologic Design for Urban Drainage and Flood Control. In *Handbook of Hydrology.* D.R. Maidment (Ed.), McGraw-Hill, Inc., New York, N.Y.

U.S. Environmental Protection Agency (1971) *Stormwater Management Model.* EPA 11024D0OC 07/71 to 1102D0C10/71, Washington, D.C.

U.S. Environmental Protection Agency (1978) *Report to Congress on Control of Combined Sewer Overflow in the United States.* EPA-430/9-78-006, Washington, D.C.

U.S. Environmental Protection Agency (1983) *Results of the Nationwide Urban Runoff Program.* Volume I—Final Report, Water Plann. Div., Washington, D.C.

U.S. Environmental Protection Agency (1984) *Report to Congress: Nonpoint Source Pollution in the U.S. Environmental Protection Agency, Washington, DC.* Office Water Program Oper., Synectics Group, Inc., Washington, D.C.

U.S. Environment Protection Agency (1992) Framework for Ecological Risk Assessment Risk Assessment Forum, Washington, D.C.

U.S. Environmental Protection Agency (1993) *A Commitment to Watershed Protection: A Review of the Clean Lakes Program.* EPA-841/R-93-001, Office of Wetlands, Oceans and Watersheds, Washington, D.C.

Waller, W.T., *et al.* (1995) Biological and Chemical Testing in Stormwater. In *Stormwater NPDES Related Monitoring Needs.* H.C. Torno *et al.* (Eds.), Am. Soc. Civ. Eng., New York, N.Y.

Warren-Hicks, W., *et al.* (1988) Ecological Assessments of Hazardous Waste Sites: A Field and Laboratory Reference Document. Report prepared by Kilkelly Environmental Associates, Raleigh, N.C.

Westerström, G. (1990) Pilot Study of the Chemical Characteristics of Urban Snow Meltwater. *Proc. 5th Int. Conf. Urban Storm Drainage.,* Osaka Univ., Jpn., 305.

Whipple, W., *et al.* (1978) Effect of Storm Frequency on Pollution from Urban Runoff. *J. Water Pollut. Control Fed.,* **50,** 974.

Young, G.K., *et al.* (1996) Evaluation and Management of Highway Runoff Water Quality. FHWA-PD-96-032., U.S. Dep. Transportation, Fed. Highw. Adm., Office Environ. Plann., Washington, D.C.

Yousef, Y.A., *et al.* (1985) Consequential Species of Heavy Metals. State of Fla. Dep. Trans. Environ. Res. Rep. No. FWHA/FL/BMR-85-286, Bureau Mater. Res., Gainesville, Fla.

Zariello, P. (1990) Seasonal Water Quality Trends in an Urbanizing Watershed in Upstate New York, USA. Paper presented at Int. Conf. Urban Hydrol. Under Wintry Conditions, Navik, Nor.

Zison, S.W. (1980) *Sediment-Pollutant Relationships in Runoff from Selected Agricultural, Suburban, and Urban Watersheds.* EPA-600/3-80-022, Environ. Res. Lab., Athens, Ga.

Chapter 2
Developing Municipal Stormwater Management Programs

Historically, municipal management of stormwater has focused on runoff quantity—primarily through the control of peak runoff rates during land development to limit susceptibility to flood-related damage. However, considerable emphasis has also been placed on municipal management of stormwater quality—pollutant removal from stormwater before discharge to receiving waters. This emphasis primarily is because of

- Regulations prepared by the U.S. Environmental Protection Agency (U.S. EPA) that control the quality of stormwater discharges under the National Pollutant Discharge Elimination System (NPDES) permitting program;
- Water quality regulatory programs at local and state levels; and
- Specific cases in which there is a clear mandate to treat stormwater before discharge, regardless of whether there is a regulatory need to do so.

Although there is legitimate debate on many aspects of stormwater quality management, there is a general consensus that urban stormwater quality needs to be managed. This section presents a six-step decisionmaking process for developing a stormwater management program, as follows:

- Step 1—define objectives,
- Step 2—assess existing conditions,

- Step 3—establish program framework,
- Step 4—select near-term best management practices,
- Step 5—implement near-term program, and
- Step 6—evaluate effectiveness.

The first three steps establish a long-range strategy for the stormwater management program. These steps are described below:

- Establish stormwater management objectives—convene a committee of "stakeholders" in the program to define what the program aims to achieve and the critical factors that will determine its success.
- Assess pollutant sources and existing controls—use readily available information to perform an initial assessment of factors affecting stormwater quality management (for example, development patterns, known pollutant sources, and observations of illicit discharges). Implement a long-term assessment plan to provide decisionmakers with a better understanding of water quality issues, pollutant sources, and best management practice (BMP) effectiveness.
- Establish a management program framework—a long-term strategy should include a management program framework outlining the scope of the program and a list of priorities appropriate to the objectives and current conditions.

Steps 4 and 5 involve selection and implementation of a complementary and integrated set of BMPs. Chapters 4 and 5 of this publication describe BMPs.

The last step in this decision process, assessing the effectiveness of implemented BMPs, may involve monitoring, modeling, and BMP performance auditing. This step is introduced in this chapter, with evaluation techniques defined in Chapter 3.

ESTABLISHING STORMWATER MANAGEMENT OBJECTIVES

An effective management program should focus on desired beneficial uses of receiving waters and must relate to significant pollutant sources within the entire watershed of that receiving water. Often, the cause–effect relationships of pollution control and beneficial use protection are uncertain. Therefore, near-term goals should focus on obvious, localized problems and should establish an appropriate scientific and administrative structure for addressing long-term beneficial use protection in a cost-effective manner.

PROTECTING BENEFICIAL USES. A stormwater quality control program should protect beneficial uses of receiving waters. These beneficial uses typically include aesthetic resources, aquatic habitat, water supply, and recreation. While state regulatory agencies have designated beneficial uses for most major receiving waters, beneficial uses of many urban streams and lakes may not be currently designated. In addition, beneficial uses of urban streams and lakes should reflect their municipal drainage functions and consider any previous or required future alterations for flood or stream bank erosion control.

After being designated, the stormwater management program should implement only those stormwater management measures necessary to mitigate urban runoff effects on beneficial uses. While this conclusion is obvious, there are significant gaps in our understanding of these effects, their causes, the relative effects of other pollutant sources, and the effectiveness of available management measures. The municipal stormwater management program should participate in confirming or designating beneficial uses of waters receiving urban runoff. Each receiving water "user" will have definite views on what these beneficial uses should be (for example, swimming, surfing, fishing, boating, aquatic life habitats, or aesthetics) and the local areas of concern. Certain waters will be seen as more valuable than others.

The attainment of seemingly obvious beneficial uses may carry significant cost burdens. If the attainment of goal 4 in Table 2.1 were selected as the goal of the stormwater management program, enormous costs would be incurred and current technology might be insufficient to meet existing standards for several constituents. Therefore, it is imperative to understand the implications of various goals and the costs associated with obtaining these goals. Technical studies are necessary to determine whether beneficial use attainment is achievable at a reasonable cost and to quantify long-term pollution control goals necessary to attain designated uses. Detailed discussion of appropriate study techniques is found in Chapter 3.

DEVELOPING A WATERSHEDWIDE APPROACH. The need to use a watershedwide approach to managing stormwater is obvious—stormwater flows and constituent loads at any location are generated to some degree by all points tributary to that location. However, every watershed is composed

Table 2.1 Beneficial use attainment guide.

	Performance standards	
Attainment goals	**Loadings**	**Water quality**
1. No significant degradation	Reduce increase	Reduce deterioration
2. No degradation	No increase	No deterioration
3. Improved water quality	Lower than existing	Better than existing
4. Meet numeric water quality standards during storm events	Significantly lower than existing	Better than existing

of many smaller watersheds, and each watershed is a portion of a larger watershed. Therefore, focus management programs on the watersheds of receiving waters with observable beneficial use impairment or valuable beneficial uses warranting protection.

Next, establish watershed management responsibilities. Smaller watersheds may be entirely urban with no significant sources of pollution other than stormwater. Typically, these watersheds can be managed exclusively by the urban stormwater management agency. However, other watersheds may span several jurisdictions or contain a wide variety of pollutant sources. Watershed management in these larger watersheds requires intergovernmental coordination and cooperation to establish

- Objectives that will benefit the watershed,
- Measurable criteria to demonstrate that the implemented practices are effective, and
- Clear lines of responsibility for watershed management among all jurisdictions in the watershed.

Finally, establish watershed management objectives. At a minimum, objectives for a small urban watershed should address the needs of its local receiving water. In addition, a small watershed's objectives should reflect its relative contribution to beneficial use impairment to receiving waters of the larger watersheds. The larger watershed may have different pollution control objectives than its smaller component watersheds, objectives that may be difficult to define in the near term because of size and complexity. Therefore, near-term stormwater management programs typically should focus on managing the watersheds of small urban streams because their needs are easier to define and solve. Near-term stormwater management programs should also address impairment to larger receiving waters only where the source is obvious, the cost is low, or the program establishes the foundation for long-term control (for example, public education).

INVOLVING STAKEHOLDERS. A stormwater management program should be developed with the involvement of one or more qualified committees of interested and affected organizations, agencies, elected officials, or citizens within the watershed. The purpose of these committees is to address technical, monetary, and policy issues associated with a regional stormwater management program. It is important to establish these committees early in the planning process. Perhaps more importantly, the committee will develop a sense of the commitment on the part of the participating agencies.

Participating committee members must include any agencies that will be affected by the stormwater management program. The composition of a typical committee is shown in Figure 2.1. Key members of the committee include representatives from regulatory agencies and environmental groups to address health and safety aspects of specific stormwater management early in

Planning Committee
Members

Lead Agency

Participating Agencies

Regional Council of Governments

Stormwater Management Program Committee

Financial Consultant

Regulatory/ Health

Technical Consultant

Legal/Policy Consultant

Figure 2.1 Stormwater management program.

the planning process. The committee also must include experienced individuals in the areas of aquatic chemistry, aquatic ecology, and urban hydrology.

The stormwater planning committees must ensure that watershed planning is conducted in the spirit of cooperation and good faith. The priorities of each participating agency need to be fully disclosed and discussed. Even a hint of a "hidden agenda" can prolong any real progress of a project.

MEETING REGULATIONS. In the final analysis, the stormwater management program pollution reduction goals are set by the regulatory requirements. This section describes types of regulatory programs concerned with urban runoff pollution.

Federal. *NATIONAL POLLUTANT DISCHARGE ELIMINATION SYSTEM PERMITTING.* Section 402(p) of the Clean Water Act established NPDES permitting requirements for municipal stormwater systems and two goals for these permits:

- Eliminate nonstormwater discharges from storm drains. Nonstormwater discharges typically are intermittent, unpredictable, widely dispersed, and difficult to detect. They may be physically connected to storm drains or may result from dry weather water use that entrains constituents as it runs off.
- Reduce stormwater pollution to the maximum extent practicable. According to U.S. EPA NPDES permitting regulations in the Code of Federal Regulations, the local entities and regulatory agency should define the maximum extent practicable (MEP) through a comprehensive planning process. The process should consider the magnitude of

the problem, constraints on its resolution, the effectiveness and track record of available BMPs, and costs. Recent practice is to define MEP using a combination of source control BMPs and treatment control BMPs. The MEP is not a fixed target but should be reexamined and, if necessary, revised to incorporate current knowledge about sources of pollution and the demonstrated effectiveness of BMPs.

COASTAL NONPOINT POLLUTION CONTROL PROGRAM. The Coastal Zone Act Reauthorization Amendments of 1990 require states with approved coastal zone management programs to develop a Coastal Nonpoint Pollution Control Program for approval by U.S. EPA and the National Oceanographic and Atmospheric Administration. The purpose of the program is to work closely with other state and local authorities to develop and implement management measures for nonpoint source pollution (including urban runoff) that restore and protect coastal waters. Coastal stormwater control programs are not intended to supplant existing coastal zone management programs. Rather, they are to serve as an update and expansion of existing nonpoint source management programs and are to be coordinated closely with the existing nonpoint source management programs (U.S. EPA, 1991).

NATIONAL ESTUARY PROGRAM. U.S. EPA administers the National Estuary Program under Section 320 of the Clean Water Act. This program focuses on all pollutant sources in geographically targeted, high-priority estuarine waters. Through this program, U.S. EPA assists state, regional, and local governments in the development of comprehensive management plans that recommend priority corrective actions to restore estuarine water quality, fish populations, and other designated uses of the water (U.S. EPA, 1991).

State. States have evolved a variety of statewide stormwater pollution control strategies. A few states initiated controls in response to the perceived needs of their states. Many more developed strategies in response to U.S. EPA requirements that have sometimes lapsed into oblivion, lacking any obvious immediate necessity or federal sanction. The general objective of controlling stormwater pollution is accepted as important not only by environmental interests but by water resource analysts as well. Development of stormwater pollution control programs is proceeding but is moving slowly.

CASE STUDY—STATE OF MARYLAND. Maryland has a well-developed program for the control of stormwater pollution, with particular interest in the restoration and protection of the Chesapeake Bay and its tributaries. Eutrophication of the Chesapeake Bay causes serious declines in water quality and fishery productivity. A water quality model of the Chesapeake Bay demonstrated the relationship between nutrient input, eutrophication, and deoxygenation. Accordingly, a target was set in 1987 to reduce the total input of nitrogen and phosphorus by 40% by the year 2000. A current reevaluation will examine the progress made toward the 40% goal and recommend adjustments to the program.

Maryland enacted legislation from 1984 to 1986 to establish a 305-m (1 000-ft) critical area around the edge of the bay and create the Chesapeake Bay Critical Area Commission to establish criteria for development in the critical area. Within intensely developed areas of the critical area, storm runoff pollution loads for new development must be reduced at least 10% below predevelopment levels or offsets must be provided. In limited development areas, no more than 20% of woodland or forest may be developed, and the impervious area created in any development site is limited to a maximum of 15% of the total site. In resource conservation areas, additional residential development can only be allowed at a density not exceeding one house per 8 ha (20 ac). State grants have been available to support the necessary local planning.

More than 5 000 stormwater management facilities have been installed on new developments in Maryland between 1985 and 1989. The general goal of Maryland's stormwater management program is that both the quantity and quality of runoff from developed land will be as close as possible to the runoff characteristics of the predeveloped condition. Implementation of management stormwater controls for existing urban areas is carried out with local jurisdictions through the Stormwater Pollution Control Program. This state program provides grants of up to 75% of the project's cost to counties and towns. Although more than 50 projects have been constructed, progress statewide has been slow.

Interstate. Many watersheds encompass areas within more than one state. For example, the Delaware Estuary Management Plan, sponsored by U.S. EPA and the states of Delaware, New Jersey, and Pennsylvania, is part of the National Estuaries Program. The Delaware Estuary, as a whole, is still relatively unpolluted. Thus, the plan focuses primarily on land use management programs in each state designed to prevent the major increases of stormwater pollution that otherwise would accompany future development and the protection of sensitive habitat (for example, a primary national flyway).

The plan establishes a "primary zone of influence" where controls would be mandated based on time of travel (that is, the time for runoff to reach the estuary). A time of travel of 40 hours at mean flow conditions was used to establish the tentative boundary for the primary zone of influence.

The estuary program is bringing attention to some striking differences in handling stormwater pollution control in the three states. Delaware has a recently developed program requiring water quality control of storm runoff from new developments statewide. New Jersey has well-developed, but nonmandatory, standards of water quality control in stormwater management in which application is left to municipal discretion. Pennsylvania has no such standards except for streams draining into lakes and for the control of nutrients in the Susquehanna River Basin related to the Chesapeake Bay estuary study. Clearly, interstate programs present numerous implementation issues.

Regional. Several states have adopted regional approaches to managing stormwater runoff. In Florida, rapid growth during the late 1970s and throughout the 1980s placed numerous demands, both in water quantity and quality, on the state's vulnerable and limited water resources. To reduce degradation from the rapid growth and extensive land-use changes, a wide variety of laws and regulations have been implemented by the state and local governments.

California's Porter-Cologne Act requires the development of basin plans for the various watersheds. The basin plans are implemented through the NPDES permitting program, which is administered by the nine regional water quality control boards under supervision of the State Water Resources Control Board. Areawide municipal stormwater NPDES permits have been granted to 12 urban counties and all cities within these counties. In the San Francisco Bay watershed, the Bay Area Stormwater Management Agencies provide watershedwide coordination of the NPDES-permitted stormwater management programs, other nonpermitted municipal stormwater management programs, and San Francisco's combined sewer overflow program. Additionally, statewide Stormwater Quality Management Task Force meetings of municipalities, industries, regulatory agencies, and consultants help coordinate stormwater management programs throughout the state.

ASSESSING EXISTING CONDITIONS

A common question asked while developing a stormwater management program is "What is the stormwater quality problem?" While this is a logical question, a firm scientific answer is sometimes difficult and costly to give. To move the program forward, municipalities should conduct a quick assessment of obvious local receiving water problems, significant pollutant sources, and existing control programs based on visual observations and readily available data. This assessment should apply monitoring, modeling, and performance auditing methods discussed in Chapter 3 for watershedwide evaluations of pollutants and their effects. Local staff members with pragmatic knowledge of the "lay of the land" are an asset when assessing existing conditions. The four primary areas that should be explored when assessing existing conditions are receiving water characterization, watershed characteristics, pollutant sources, and existing programs.

RECEIVING WATER CHARACTERIZATION. The main receiving water effects associated with stormwater are

- Receiving water hydromodification because of increased runoff from urban areas;
- Water quality degradation from changes in the chemical/bacteriological constituents in runoff; and
- Losses in the assimilative capacity of the urban drainage system.

Schueler (1993) classifies urban runoff effects as changes in stream hydrology, morphology, water quality, habitat, and ecology. An effective stormwater management program needs to consider receiving water characteristics and effects of urban runoff on the receiving water. However, significant gaps remain in our understanding of and ability to define these characteristics and effects. Therefore, the near-term stormwater management program should begin to define receiving water characteristics and effects to establish near-term stormwater pollution control targets.

Typically, the beneficial use effects of stormwater on specific receiving waters have not been examined in detail. The traditional method of defining beneficial use effects are based on numerical and narrative water quality standards and effluent limits for chemical constituents. This traditional method must be supplemented by other approaches (for example, hydrologic modeling, bioassessment, and habitat assessment) to address the full range of urban runoff effects and develop appropriate control strategies. Chapter 3 discusses these traditional and alternative approaches to characterizing receiving waters and the effects of urban runoff.

POLLUTANT SOURCES. The stormwater management program should direct control measures at pollutants that impair the beneficial uses of the receiving water. For example, pollution control in the watershed of a eutrophic receiving water should be directed at significant sources of the limiting nutrient (such as phosphorus or nitrogen). Concentrated sources of pollutants can be effective early targets for the stormwater program because their cleanup is both observable and measurable. These sources might include old commercial and industrial areas and old core urban areas where polluting non-stormwater connections to the storm drain system are most likely to be found. Also, stormwater pollution controls are often easier to implement if directed at related problems recognizable to the community. For example, a business or resident using "bad housekeeping practices" outdoors is often considered to be an eyesore to neighbors (if not in violation of local regulations).

Visual observation is the best way to begin targeting pollutant sources in the near term. After the most concentrated sources are addressed, long-term monitoring programs may begin providing insight to more widely dispersed pollutant sources. Computer models then can be used to project existing and future pollutant loads based on the land uses of the watershed and characteristic concentrations of runoff from each land use (see Chapter 3 for details). Research into the composition of materials may also help reveal the source of critical pollutants.

Pollutant load estimates define the relative loading of pollutants from different parts of the watershed. Knowledge of pollutant characteristics helps reveal proper pollution control strategies. For example, can nonstructural controls, especially for currently existing development, significantly address water quality needs before structural controls are even considered? What will be gained if structural controls are retrofit in an area, and can the cost be justified?

WATERSHED CHARACTERISTICS. The basic objectives of this step include becoming familiar with the watershed, focusing on those factors that will influence the nature of stormwater management, and collecting all information necessary to select BMPs. Characteristics should include the following:

- Land use: What is the existing land use and impervious area of the watershed? What will land use/imperviousness be at buildout? What is the proximity of intensive land uses to sensitive receiving waters?
- Physiology: Are slopes steep, moderate, or flat? Are soils erosive? Are they permeable? What is the depth to bedrock? To groundwater?
- Climate: What is the rainfall intensity/duration/frequency for small storms?
- Habitat: What habitat thrive in this area? How has urbanization irrevocably altered this habitat?
- Drainage system: What impervious surfaces are connected to the drainage system? What is the extent of the closed storm drain system? Where do curb and gutter systems exist? Are open storm drains improved or natural?
- Community profile: What are the makeup and activities of the residents and businesses in the watershed (for example, demographics, community organizations, and business climate)?

EXISTING PROGRAMS. An effective stormwater management program consists of a number of specific components, each targeted at specific types of pollutant sources. However, not all components need to be new components focused exclusively on stormwater management. Every municipality conducts a wide variety of programs that can be integrated to a community-wide stormwater management program. This section presents examples of existing municipal programs that have objectives similar to those of stormwater management.

Existing Regulatory Programs. Table 2.2 lists a number of regulatory programs that have a bearing on the selection of BMPs or implementation of a stormwater management program. The prudent selection of BMPs will serve to comply with the objectives of both the stormwater program and other regulatory programs. The following paragraphs describe various federal, state, and local programs as they relate to stormwater control.

Table 2.2 Existing regulations for stormwater pollution control.

Regulation	Activity	Potential
Federal Clean Water Act 401 and 404 permits	Permits dredging and filling in "waters of the United States"	Erosion control, sediment control, long-term sediment balance, and reduce pollutants Vegetative controls to preserve riparian areas
Federal 1601 and 1603 Stream Bed	Alterations to creek and stream beds	Pollutant controls and prevent loss of habitat
State General Plan Act	Municipal development objectives Adoption of ordinances	Stormwater management controls, for example, pollutants control, watershed protection
State Environmental Quality Act	Environmental review of projects	Mitigation measures for reduction of pollutants
State Subdivision Map Act	Adoption of ordinances	Standard/regulations for grading, erosion protection, detention/retention design, and dust control
State air quality management plans	Emission	Sediment and dust controls
Local flood plain management and drainage ordinances	Control of velocity Detention/retention Bank stabilization and outlet controls	Control of erosion Control of sediment, pollutants, and quantity Erosion and sediment controls
Local zoning ordinances	Cluster development Hillside development Landscape/open space	Reduce runoff and impervious areas Slope and erosion restriction; may include revegetation or stabilization Vegetative best management practice perimeter controls reduction of runoff
Local sewer use ordinance	Control of illicit connections	Pollutant controls
Local uniform building code	Chapter 70—excavating and grading	Reduce erosion and sedimentation Standards for stable cut and fill slopes
Local uniform plumbing code	Prevention of illicit connection Various chapters on materials and application/use	Pollutant controls Pollutant controls
Local fire code	Storage of materials	Pollutant controls

PESTICIDE PROGRAM. U.S. EPA administers the pesticide program under the Federal Insecticide, Fungicide, and Rodenticide Act. Among other things, this program authorizes U.S. EPA to control pesticides (sometimes found in stormwater) that may threaten groundwater and surface water (U.S. EPA, 1991). Potential actions carried out in this program include national requirements on labels, training, development of state management plans, and national prohibition of certain domestic uses of designated chemicals.

HAZARDOUS MATERIAL/WASTE CONTROL PROGRAMS. There are numerous laws (primarily the Resource Conservation and Recovery Act and the Emergency Planning and Community Right-to-Know Act) and regulations regarding the control of hazardous material and waste. Hazardous materials storage and emergency response programs regulate hazardous materials storage and emergency response planning. A company's annual business plan must include a hazardous material inventory, estimates of hazardous waste amounts, and emergency response planning. Under workers' right-to know programs, the employee is advised, through the use of material safety data sheets, material labeling, and employee training, of the potential for contact with hazardous substances. Also, under hazardous waste source reduction and management review programs, hazardous waste generators must look at source reduction as the preferred method for managing waste. The industry must prepare a source reduction evaluation review and plan that identifies all hazardous waste streams and potentially viable source reduction approaches.

AIR QUALITY PROGRAMS. Source control of atmospheric contributions of stormwater pollution (such as automobile and industrial emissions) should be coordinated with state and local air quality programs. As an example, some states require counties with a metropolitan population greater than 100 000 to form a Congestion Management Agency (CMA). This agency, which typically is a joint power authority, coordinates the development of a Congestion Management Program. This program addresses the effects of land-use decisions on regional transportation systems, trip reduction ordinances, and public transit services. Consequently, close cooperation between the CMA and the municipality would benefit each entity's efforts to reduce pollutants in the environment.

SPILL PREVENTION AND CLEANUP PLANS. Federal regulations require on-shore facilities engaged in operations that could reasonably be expected to discharge oil in harmful quantities to prepare Spill Prevention Control and Countermeasure (SPCC) Plans. National Pollutant Discharge Elimination System regulations for some industrial activities also require SPCC plans as part of the facility's BMP program.

Existing Municipal Programs. An effective stormwater management program should instill stormwater management concepts to the activities of other municipal programs. Several examples of integrating stormwater management to existing programs follow:

- Integrated pollution prevention programs–in many cases, a single pollutant prevention measure can meet more than one environmental regulation (for example, household hazardous waste collection, waste minimization, landfill management, or pesticide use).

- Integrated water resource management–extending flood control policies to address more frequent storms, prevent streambank erosion, and promote percolation of runoff to control the peak and volume of runoff will also control pollutants in runoff.
- Growth management policies–how and where land is developed may significantly change the effect of stormwater pollution on receiving waters. Directing intensive land uses away from sensitive receiving waters has been proven to be an effective source control BMP for new development projects.
- Multiuse open spaces–many communities incorporate treatment control BMPs to parks and recreation areas through proper planning, design, and landscaping. These types of "multiuse" facilities provide a cost-effective way of meeting stormwater management plan goals.

ESTABLISHING A MANAGEMENT PROGRAM FRAMEWORK

STRUCTURING THE PROGRAM. A stormwater management program must address a wide variety of individuals, businesses, organizations, and agencies that contribute to the stormwater pollution problem and its solution. Such a program should be planned carefully to have the following important characteristics:

- Comprehensive—the program should have adequate resources to address all aspects of stormwater quantity and quality in the watershed that affect beneficial uses of receiving waters.
- Integrated—maximum effectiveness is achieved by an integrated program of regulation, education, and municipal action.
- Balanced—each responsible party must be defined, and specific direction must be given to each party. Direction for each party should be commensurate with its share of the problem and equitable to its share of the solution.
- Continuous and dynamic—meaningful stormwater control, particularly in developed areas, will take a long time, certainly longer than a single 5-year NPDES permit term. Therefore, stormwater management programs should be implemented in a manner that addresses the dynamics of an evolving, long-term program.

The program's organizational framework should capture the full breadth of the problem and its solution. It is not appropriate for this publication to recommend an ideal stormwater management program structure. Each program must do this based on the assessment of the existing conditions described in the previous section. Typically, however, each program should address the three broad areas listed below:

- Land development is one of the most typically regulated aspects of stormwater management in North America and, thus, is the area where the most experience lies. This aspect of the program typically addresses land-use planning considerations, incorporating treatment controls to development projects, and controlling construction-related effects.
- Municipal drainage system management typically involves cleaning channels, detention basins, pipes, inlets, catch basins, streets, and impervious areas connected to the drainage system. In some cases, it may be feasible to retrofit treatment controls to the drainage system to improve the ability to capture stormwater pollutants or ease their removal.
- Residents and businesses also contribute stormwater pollution, largely by conducting outdoor activities in a manner that allows pollutants to enter the drainage system. Municipalities can use a combination of education and regulation to address this aspect of stormwater pollution.

SETTING PRIORITIES. The next step in developing and implementing the stormwater management program is to set priorities and establish phases for meeting the established objectives. Priorities and phasing should be soundly based on the existing condition assessment. Priorities should be established by the committees that have defined the program objectives. Many decision-making techniques exist for establishing priorities. It is not the intent of this publication to discuss these techniques. At a minimum, priorities should reflect common sense, using questions like those listed below:

- How much can we afford?
- Do we think it will work?
- What are we already doing?
- Can we show that it works?
- Where are the worst problems?
- Can we learn something new?
- What are the easiest solutions?
- Is this the right direction?
- What has been successful for others?
- Will beneficial use impairment decrease?

SELECTING MANAGEMENT PRACTICES

The basic objective of stormwater management is watershedwide improvement in water quality and enhanced beneficial use of the receiving water bodies. The current practice is to presume that a cost-effective, practicable set of BMPs (that is, restrictions, techniques, or treatment facilities that are required under given conditions) will provide some progress in protecting water quality. For example, individuals can conclude that used motor oil should be recycled rather than discarded without site-by-site analysis or sampling. This approach typically will be used, at least in the near term, because the cause–effect relationships between pollutants in the watershed and beneficial use impairment in the receiving water are not always understood, the urban stormwater management agency often does not have complete jurisdiction over a complete watershed, or regulatory requirements can be met through BMPs.

MULTILEVEL STORMWATER QUALITY MANAGEMENT STRATEGY. A municipality's stormwater management strategy must cost-effectively address program objectives using existing information to address priority needs within the context of an overall program framework. An effective strategy uses multiple BMPs, including source controls (see Chapter 4) and treatment controls (see Chapter 5). A single BMP typically cannot provide significant reductions in stormwater pollutant loads because these pollutants come from many sources within the municipality. However, multiple BMPs can provide complementary water quality enhancement to achieve desired results. A multilevel BMP approach, schematically depicted in Figure 2.2, deals with the many pollutant and runoff sources throughout the watershed and shows that, whenever feasible, combining most effective BMPs in a series can be an effective strategy to reduce pollutant loads being transported to the receiving waters by stormwater (UDFCD, 1992).

SELECTING STANDARD AND SPECIAL BEST MANAGEMENT PRACTICES. There are hundreds of different BMPs appropriate for the different sources of pollution and the varieties of receiving waters to be protected. In practice, a large degree of simplification will be necessary, especially in starting a program. Typically, there should be a set of standard BMPs appropriate for general application to a particular geographical or political area, supplemented by special BMPs for application where a greater degree of pollution control may be desirable. These would be further refined by establishing standard and special BMPs for the protection of groundwater and surface water, respectively. For example, control of nitrates and volatile organics is more important for groundwater than for rivers unless they drain into a reservoir, lake, or estuary. Therefore, special BMPs for those substances might be applicable for percolation to groundwater but not for drainage to rivers. In addition to specifying either standard or special degrees of protection, additional measures may be added as particular circumstances require, and, of course, there are some cases where no control of stormwater pollution is necessary.

It is apparent that the selection of BMPs must consider the potential harmfulness of the land use, the sensitivity of the receiving waters, and the prox-

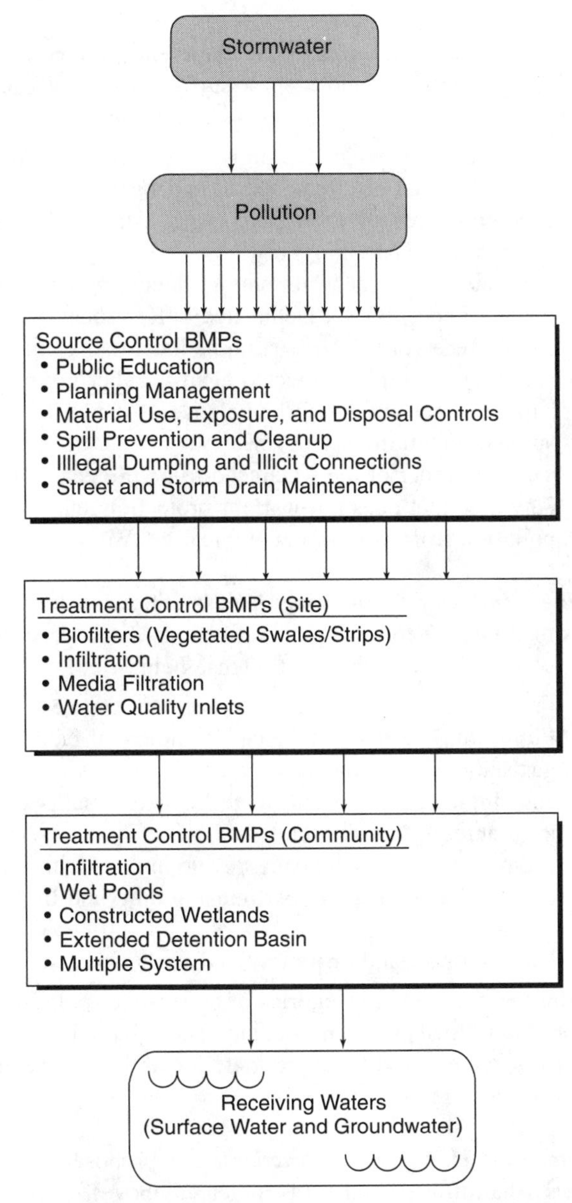

Figure 2.2 Multilevel strategy for stormwater management program (UDFCD, 1992).

imity of the land use to the sensitive portion of the receiving water. One BMP selection method uses numerical indices of harmfulness and sensitivity (see Figure 2.3). These indices can be evaluated with a matrix approach to show whether special or standard BMPs will be required for a given situation or whether, through land-use controls, that kind of development should be prohibited in the location proposed. However, best results can be obtained if the approach described previously provides for the level of BMP control

Areal Priorities: Designate areas GW where extra protection is warranted for groundwater and SW for surface waters warranting extra protection:

Class 1 Any officially designated highest priority protection areas, such as habitat of threatened or endangered species, exceptional resource wetlands. Also, delineated well head protection areas (GW).

Class 2 Aquifers usable for water supply, including important recharge areas (GW). Delineated buffer zones around water supply reservoirs. The area immediately adjacent to the reservoir should be protected against either groundwater or surface water pollution, the balance of the buffer against surface pollution only (SW).

Class 3 The remainder of the watershed to be protected.

Class 4 Any area that does not warrant protection against water pollution from stormwater (typically GW).

Harmfulness Index of Stormwater: Unless otherwise demonstrated in specific situations, the following classifications are in descending order of pollutant loading, the most intense being first:

Class 1 From industrial and waste management sources.

Class 2 From multiple-family housing, commercial facilities such as gas stations and shopping centers, highways, urban areas, and single-family housing with lot sizes smaller than one-third acre per housing unit.

Class 3 From single-family housing developments, with lot size one-third acre or larger per housing unit, and from lesser roads.

Class 4 Undeveloped land or unfertilized vegetation.

In addition, there are other categories of stormwater pollution that are highly variable in their pollutant loading. These include agriculture and road salts, which must be appropriately classified depending on the circumstances.

Stormwater Best Management Practices: If a proposed land use is rated Class 1 harmfulness index (as indicated above), it should not be located in a Class 1 area. If located in a Class 2 or 3 area, special provisions for quality control of runoff are appropriate. If the proposed development is Class 2 harmfulness index, special BMPs are appropriate if the development is located in a priority 1 area. (Class 1 developments may not be allowed in such areas.) If the proposed Class 2 development is not located in a priority 1 area, but its runoff drains into it, it may still require special BMPs for runoff control. A decision is required as to how far upstream such controls should extend. Otherwise, standard BMPs are appropriate for Class 2 harmfulness, except for priority 4 areas.

Figure 2.3 Best management practice numerical selection guide (ac × 4 047 = m^2).

Infiltration Best Management Practices:

a. In Class 1 protected areas, such as well heads, buffer areas adjacent to protected habitat, and habitat of endangered species, facilities involving infiltration may not be allowed at all. If they are allowed, infiltration of Class 1 runoff will be acceptable only with special water quality provisions for processing the runoff. Infiltration of processed Class 2 runoff will be acceptable if processing of runoff follows standard BMPs.

b. In Class 2 areas, including aquifers, Class 1 runoff that has been processed through standard water quality systems such as water quality detention basins or filter strips may be approved for infiltration to aquifers. Infiltration of Class 3 runoff should be encouraged, after processing with standard water quality provisions, unless special circumstances indicate existence of potential contamination.

Conditions for Infiltration

Areal Priority	Harmfulness Class 1	Harmfulness Class 2
1	No infiltration except after special water quality process	No infiltration except after standard water quality processing
2	No infiltration except with a positive showing of no adverse effect or after standard water quality processing	Infiltration encouraged unless potential harm indicated
3	Allowed	Allowed
4	Allowed	Allowed

Figure 2.3 Best management practice numerical selection guide ($ac \times 4\,047 = m^2$) (continued).

needed for an entire watershed and addresses BMPs that can be applied throughout the watershed. A detailed watershed-planning process should be used where the type and level of BMPs at each site may differ because of technical, financial, or jurisdictional issues.

SELECTING BEST MANAGEMENT PRACTICES FOR LAND DEVELOPMENT. Land development changes both the hydrologic regime of a watershed and the chemical constituents in the runoff. These changes are best controlled at the point of origin by land-use planning. Next best are source controls because land-use controls are often difficult to implement. To the extent that land-use controls and source controls are not sufficiently effective to prevent pollution, treatment controls should be used to address the problem. These controls may include detention and retention facilities, other measures classified as treatment, or, occasionally, diversion of runoff to a less sensitive environment. Both source controls and treatment controls may be included as conditions of a permit, as indicated further below.

Land-Use Controls. Land-use controls involve adoption of a comprehensive and integrated set of environmental restrictions to govern the development process. The greatest level of beneficial use protection is afforded when a single development ordinance is adopted by a community and administered by a single planning authority. In short, the ordinance mandates a minimum level of environmental site planning during development and can include, but is not limited to, the following items:

- Stream buffer requirement—development is not allowed within a variable width buffer strip on each side of ephemeral and perennial stream channels. As an example, in Baltimore, Maryland, the minimum width of the buffer strip is 15 m (50 ft) for low-order headwater streams but expands to as much as 60 m (200 ft) in larger streams (Baltimore County, 1989). The stream buffer further expands to include floodplains, steep slopes, wetlands, and open space areas to form a contiguous system, according to prescribed rules.
- Floodplain restrictions—development is restricted within the boundaries of the postdevelopment, 100-year floodplain.
- Steep slope restriction—no clearing and grading are permitted on steep slopes, for example in excess of 25%.
- Nontidal wetland protection—no development is permitted within nontidal wetland areas and a perimeter buffer area.
- Protection of environmentally sensitive areas—development is not allowed within unique habitat areas/plant communities or protective perimeter buffers identified through watershed master planning.
- Upland and riparian tree cover requirements—an allotted percentage of upland predevelopment tree cover must be maintained after site development. In addition, the riparian tree cover (which should be entirely contained within the stream buffer system) must also be retained or reforested if no tree cover currently exists.
- Waterway disturbance permits—forms of development such as roads and utilities must, by their very nature, cross through the stream buffer system. Linear developments must be scrutinized to locate them in the narrowest portions of the buffer system and ensure they do not form barriers to either fish migration or riparian wildlife migration.
- Community open space requirements—after the stream buffer system has been delineated, consider preserving additional open space at the site to accommodate the residents' future requirements for parks, playgrounds, ball fields, and other community needs.
- Cluster development—the objective here is to reduce the impervious surfaces at the site and cluster development into centralized areas where stormwater can be effectively treated. The best tools include

transferable development rights, cluster zoning, site "fingerprinting," planned unit development, and flexible site and road width layout (Yaro *et al.*, 1988).

Selecting Source Controls. A source control program for a land development project should

- Identify possible postconstruction outdoor activities that may use or generate concentrated or high-risk pollutants at the site;
- Prohibit these outdoor activities where practical;
- Designate specific areas for those activities that must be conducted outdoors;
- Install structural source controls (for example, covers, enclosures, containment systems, or connections to sanitary sewers) in these designated areas; and
- Place conditions on the development project for maintaining these areas.

Outdoor activities may include material storage, waste handling, material loading or unloading, vehicle and equipment maintenance, and myriad specific work tasks typically conducted outdoors. Source controls are selected as described in Chapter 4 of this publication.

Selecting Treatment Controls. An effective system of treatment controls addresses dispersed sources of pollutants throughout the watershed that affect beneficial uses but cannot be effectively controlled at the source (for example, automobiles and air deposition). When coupled with proper land-use controls, they can be integrated to the landscaping, drainage and flood control system, and other open spaces of development projects. When properly designed, they become amenities rather than interferences to development projects. When preceded by effective source controls, they do not cause other environmental problems (such as groundwater pollution or hazardous sediments). For example,

- A grass-lined channel can be designed to effectively convey flood waters and, by incorporating certain design techniques, reduce pollutant loads.
- Ponds can be designed to attenuate peak discharges to desired levels and reduce pollutant load before discharge. Pollutants are reduced by detaining "first-flush" runoff from large storms and all runoff from small storms long enough for sediments to settle and biological processes to act on degradable materials.
- Many well-designed treatment controls integrate with recreational facilities, provide wildlife habitat and groundwater recharge, preserve open space in urban settings, and increase the value of adjoining land.

Integrating treatment controls to multiuse facilities of this type promotes cost effectiveness and other benefits by reducing capital costs (by reducing the number of facilities needed in the first place); saving land; reducing operation, maintenance, and replacement costs; reducing planning and design time; and stimulating integrated, comprehensive planning. Chapter 5 presents an integrated approach to selecting and designing treatment controls that achieves desired water quality and quantity objectives in an effective and efficient manner.

SELECTING BEST MANAGEMENT PRACTICES FOR EXISTING MUNICIPAL DRAINAGE SYSTEMS. In most cases, the urban drainage system includes street gutters, inlets, catch basins, storm drain pipes, constructed ditches, channels, and urban streams. These streams often are severely altered, either by humans or by the changed hydrologic regime. Two categories of BMPs are appropriate for urban drainage systems:

- Various techniques for cleaning drainage system components or preventing nonstormwater discharges to the drainage system in the first place (source controls); and
- Various devices that can be retrofit to the drainage system to slow the rate of runoff or remove and assimilate pollutants (treatment controls).

Typically, all available source controls for municipal drainage systems should be used to some degree. Selection of BMPs for existing municipal drainage systems involves establishing an appropriate cleaning frequency for each component and an inspection frequency for nonstormwater investigations. Chapter 4 describes the source controls for municipal drainage systems and presents guidelines for determining cleaning/inspection schedules.

Feasible retrofit strategies typically involve an urban stream restoration program and feasible retrofitting of existing flood control basins. The primary purpose of stream restoration is to enhance the aquatic habitat and ecological functions of urban streams that have been lost or degraded during the urbanization process. Stream restoration also "builds back" the assimilative capacity of these urban streams, allowing them to better remove pollutants before discharging to larger downstream receiving waters. A comprehensive stream restoration program incorporates several of the following steps, where geomorphological/land-use opportunities permit:

- Watershed assessment of restoration opportunities—stream restoration opportunities are best assessed though systematic biological surveys throughout the stream system and its upland watershed. These surveys should determine the dominant effects that have degraded the aquatic community and identify feasible opportunities for restoring stream habitat or water quality.

- Retrofitting of flood control basins—the best restoration opportunities often involve the improvement of existing flood control basins. For example, in the Washington, D.C., metropolitan area, retrofitting typically has involved converting older, dry stormwater ponds to extended wet pond–marsh systems (Herson, 1989).
- Construction of additional urban best management practices—retrofit of new urban BMPs in the urban landscape is not an easy task, given the typically limited amount of space available. Innovative retrofit techniques include the peat–sand filter (Galli, 1989), oil–grit separator inlets, and extended detention lake/wetland systems (Schueler and Helfrich, 1988).
- Reforestation programs—the buffer zone along urban streams and upland areas can be reforested gradually within a matter of years at a relatively low cost through cooperative community tree planting programs according to a long-term watershed plan. These volunteer programs are most effective when local governments arrange the logistics, assemble the sites, and secure the plant stock.
- Instream fish habitat improvement—the degradation of stream habitat structure (most notably the loss of pools, riffles, and clean spawning areas) can be reversed by adapting habitat improvement techniques such as boulder and log deflectors, log drop structures, brush bundles, willow wattles, boulder placement, and imbricated riprap.
- Urban wetland creation/restoration—active creation, restoration, and management of urban wetlands partially substitute for the lost ecological functions of the destroyed or degraded wetland system in many urban areas.
- Identification and removal of fish barriers—possible barriers to fish migration can be detected through systematic upstream/downstream fish collections at suspected structures during spring runs (Cummins, 1988) or, in some cases, by visual surveys. In many cases, urban fish barriers are created by structures in the stream that can be rather easily modified to allow migration.
- Stream stewardship—citizens often take an active and personal interest in maintaining urban stream quality. Local governments can encourage them to adopt a stream; remove debris; and report oil spills, sediment control violations, pollution problems, and sewer overflows.
- Natural urban parks—urban parks, centered about the stream valley and riparian vegetation zone, provide a tool for managing land use and human access while protecting key vegetative communities in core stream areas.

SELECTING BEST MANAGEMENT PRACTICES FOR RESIDENCES AND BUSINESSES. Most of the urban landscape consists of private property containing residences and businesses. An effective municipal stormwater management program must include BMPs for these residences and businesses. Typically, the near-term program will focus on source control in these areas, achieved through a combination of public education and municipal regulation. Chapter 4 of this publication describes feasible source control strategies for existing residences and businesses.

It is unlikely that treatment controls will be retrofit to existing development on private property in the near future unless the runoff is significantly polluted, the property discharges to a highly sensitive receiving water, source controls are ineffective, or the property is subject to an industrial NPDES stormwater permit. Retrofitting other treatment controls on private property will be expensive and controversial without additional proof of direct water quality effects for each property's runoff.

IMPLEMENTING THE PROGRAM

IMPLEMENTATION STRATEGIES. Three options exist for implementing the stormwater management program: incorporating stormwater management concepts into an existing program, starting a new program, and conducting further planning and study. The following sections describe each of these strategies.

Incorporate Stormwater Concepts to an Existing Program. Programs that partially achieve stormwater pollution control are already in place for many areas. These may be municipal programs, programs sponsored by overlapping jurisdictions, or private initiatives. Typical existing programs include household hazardous waste collection, hazardous spill response units, hazardous material storage rules, Resource Conservation and Recovery Act and landfill management and closure regulations, litter control programs, and erosion and sedimentation control requirements. For many existing programs, all that may be required is better documentation and effective communication about their conduct and effectiveness at achieving stormwater pollution control. Other existing programs may be compatible with stormwater pollution control but require some redirection. In this category, the objective should be to incorporate effective stormwater management concepts to the program. Water quality management plans that have been developed by local governments to satisfy the requirements of Section 208 of the Clear Water Act typically refer to nonpoint source effects and controls. These need to be reviewed and, in most cases, redirected to focus on stormwater quality aspects rather than on point sources only. Existing programs that may be redirected include detention/retention requirements for new development, construction site and building permit inspection programs, drainage system maintenance activities, and transportation and land-use planning programs.

Start a New Program. There are two instances where new programs are favored in the near future:

- The program will directly and measurably benefit the beneficial uses of the receiving water (for example, an illicit discharge control program).
- The program will begin educating the community at large about how stormwater systems operate, how their actions can pollute receiving waters through storm drains, and how the community can be involved in stormwater pollution control.

Conduct Further Planning and Study. Some activities warrant additional study in the short term to achieve cost-effective BMP selection:

- Watershedwide master plans are often used to select BMPs for land development activities and new, enlarged flood control facilities as described previously in this chapter.
- Pilot studies are an interim step aimed at gaining practical knowledge on which to base long-term implementation decisions. They should be directed toward a representative microcosm of the watershed and should be considered where watershedwide implementation is expected to be cost prohibitive in the near term. Possible pilot studies may include street sweeping programs, school curriculum on stormwater pollution control, stormwater pollution control initiatives directed at small businesses, innovative treatment technologies, and illicit connection detection programs.

Pilot studies are most appropriate for source control BMPs where a high level of community involvement may be required, implementation may be labor intensive, effectiveness is largely unmeasured, or effective implementation may be site specific. While a large body of research and practical field experience already exist for treatment control BMPs, pilot studies of these BMPs may still be necessary to account for local climatic, physical, and institutional conditions.

TREATMENT CONTROL OPERATION AND MAINTENANCE. Regular inspection and maintenance of treatment BMPs is an absolute necessity if these controls are to perform consistently to expectations. Sediment removal and removal of debris from BMP inlets and outlets are important requirements to maintain consistent performance. Access to detention ponds is necessary for excavating equipment, trucks, mowers, and personnel for routine maintenance and erosion repair and for the removal of sediment and trash accumulation. Where access is particularly difficult or impractical, ponds should be overdesigned to allow for additional sediment accumulation.

Chapter 5 of this publication presents an overview of maintenance requirements. Further information is also contained in *Design and Construction of Urban Stormwater Management Systems* (ASCE and WEF, 1992). Each municipal stormwater management program should determine specific maintenance practices for treatment controls implemented in their community, clearly define what the municipality will maintain, and develop enforceable agreements with private parties responsible for facilities the municipality does not maintain.

FINANCING STORMWATER MANAGEMENT PROGRAMS. Revenue Alternatives. The stormwater management program needs a reliable source of revenue for general administration, operation, maintenance, and capital improvements. This section reviews revenue alternatives for the funding and financing of the stormwater management program. Funding refers to sources of revenues used to pay for annual operating expenditures, including maintenance and administrative costs; to pay for improvements directly out of current revenues; and to repay debt issued to finance capital improvements. Financing is defined as the initial source of funds to pay for planned capital improvements or equipment.

Five funding alternatives and nine financing alternatives are reviewed in this section. Several important characteristics are discussed to provide a general background for comparing alternatives. Legal provisions for the selection and use of the revenue alternatives presented may vary from state to state, depending on state statutes or regulations.

Funding Options. Annual funding requirements of a local stormwater management program may include administration, permitting, design, planning, operation, system maintenance, meeting the system's annual debt service obligations, and other financial requirements.

GENERAL GOVERNMENT TAX RECEIPTS. Stormwater management operation and maintenance costs can be funded by general government tax receipts. With this option, the funds required could be generated by any revenue source accrued in the general fund. Revenue from the general fund primarily is derived from *ad valorem*-based levies on taxable property. A significant advantage to consumers is that these costs are tax deductible. The primary disadvantage is that the general fund is not a dedicated source, and stormwater management needs would have to compete for limited funding with other programs. As a result, this funding mechanism is subject to fluctuations and typically is unreliable for implementing a long-term capital improvement program. In addition, *ad valorem* funding is less equitable because tax-exempt property does not pay and charges are based on property values rather than the effects associated with stormwater runoff or services provided.

BENEFIT AREA OR SPECIAL ASSESSMENTS. Property owners within a stormwater benefit area typically are assessed a levy or charge to fund stormwater systems within the benefitted area. In addition, if different land uses within the benefit area receive substantially different levels of stormwater benefits, the assessment of levies from subarea to subarea should vary in proportion to the benefits received.

The following criteria typically are applied when a benefit area or special assessment is developed:

- Assessment levies should not exceed the amount of the benefit received by any particular property;
- The assessment should be properly allocated to the benefitted properties; and
- Property owners typically are allowed by law to have an opportunity to comment on how the assessments are allocated to their properties.

The goal is to show that assessed levies are used to cover costs of facilities or services benefitting each property and that the benefits to each property are at least equal in value to the assessment levy. Constitutional standards typically require that property owner benefits funded by such assessments be special benefits that typically are not shared by the community as a whole.

STORMWATER MANAGEMENT UTILITY. A dedicated funding source that has gained recent acceptance is the establishment of a stormwater management utility (SMU). The premise of this method is that developed property can be charged a user fee in proportion to the contribution to the need for stormwater facilities or services. Utility customers are the properties that add runoff to the stormwater system, and individual charges can be calculated using an appropriate billing unit formula. The charge per billing unit can be set to generate the level of annual funding desired by the community.

The careful development of a rate structure has been found to be critical to the successful implementation and operation of an SMU. Also, feasible procedures for preparing billing records and database management requirements should be defined early.

The basis for determining the fee a customer pays for stormwater management services typically is the amount of runoff generated by property improvements. The runoff potential is a function of many variables and is often best determined by measuring the area of impervious surfaces. Because all improved property, including tax-exempt property, can be charged in proportion to its effect on or need for the stormwater management system, SMU fees have been proven to be an equitable and fair method of paying for stormwater management services.

INSPECTION FEES. The costs of evaluating permit applications and inspecting ensuing construction activities can be recovered through fees assessed to the developers or applicants who create the need for this service. These fees can be used with other funding options to recover the costs of providing user-specific stormwater management services.

OTHER FUNDING. Funding mechanisms could include other taxes such as fuel or utility taxes. The availability of other options will depend on state and local statutes that must be evaluated on a case-by-case basis.

Financing Options. This section reviews options available for initial financing of stormwater system capital improvements or equipment purchases. Several of the options discussed involve the issuance of bonds, which is one approach to obtaining lump sums of revenue to pay for significant capital improvement or equipment expenditures. A bond is an instrument by which an agency borrows money and guarantees repayment of the loan principal and interest with revenues generated from a specified source.

PAY AS YOU GO. This approach seeks to establish a revenue flow that is accumulated in a sinking fund. Planned facilities would be constructed or equipment would be purchased after the sinking fund contains enough revenue to pay for the purchases. Revenue deposited in the sinking fund can be derived from a number of sources, including tax revenues, user charges, in-lieu-of charges, or system development charges (impact fees).

GENERAL OBLIGATION BONDS. A general obligation bond is a loan that is secured by the full faith and credit of a local government. This means that the local government pledges its general fund revenues, primarily property taxes, as the source of revenue for loan repayment. This type of bond is easy to administer and widely used to finance various types of capital improvements. Because of the high level of security associated with the pledge of tax revenues for these bonds, the interest rates can be lower.

Both the amount of general obligation debt that a local government may issue and the total property taxes that may be assessed are limited. Because general obligation bonds are repaid through property taxes, the distribution of costs typically is not considered equitable. Costs are distributed to property owners in proportion to the value of their property, rather than to the amount of stormwater runoff generated by their land use. Tax-exempt properties do not pay for these improvements.

REVENUE BONDS. A utility enterprise typically can fund capital improvement projects or purchases by issuing bonds that will be repaid by the revenues from an enterprise operation. With revenue bonds, the utility obtains the funds needed to start and complete significant improvements and repays them over the long term with revenue collected from those who benefit from the improvements, such as the utility's users or rate payers. Provided the utility has a rate structure designed to equitably recover costs, this form of financing is better suited for use on stormwater projects than general obligation bonds. Interest rates, thereby the interest expenses, typically are somewhat higher for revenue bonds than for general obligation bonds.

SPECIAL ASSESSMENT BONDS. Special assessment bonds may be issued if the planned stormwater construction project will benefit a specific area or watershed. These bonds are secured by non-*ad valorem* tax assessments to the landowners in that area. The initial cost of implementing this option typically is high because the local government must establish stormwater benefit areas or tax assessment districts for each individual project. This bond is then relatively easy to administer because the debt service is recovered through non-*ad valorem* assessments to the properties in the district.

User acceptance and equity can be good for a properly designed program because a clearly identified area directly benefits from the stormwater improvements. User equity also can be high because tax assessments must be distributed in direct proportion to the benefits received.

DEVELOPER-CONSTRUCTED IMPROVEMENTS. As a condition of approving a proposed new development, the local government could require a developer to construct improvements to control stormwater runoff. Typically, these are on-site improvements; however, the local government occasionally may require some off-site improvements. The ease of administering this option may be difficult because developers may object to or oppose the required construction. Equity typically is considered to be good because the new development likely to generate the runoff will incur the cost of controlling or alleviating the problems arising from that runoff.

IN-LIEU-OF CHARGES TO DEVELOPERS. An alternative to requiring developers to construct on-site stormwater management facilities is to require them to pay a front-end charge for off-site capital improvements needed to serve their development. Payment would be a condition of development approval and would recover the cost of the off-site improvements to manage the development's runoff or its proportionate share of the cost of a regional facility serving a larger area.

This technique frequently is used where a small-scale facility is not necessarily advisable (that is, assumption of responsibility for the operation and maintenance costs for the planned improvements), and the local government wants developer participation in a larger facility designed to control drainage on a regional basis.

Conceptually, this option can be equitable because it can be designed to recover the cost of the improvements in proportion to the runoff generated by the users of the system. However, equity may be difficult to implement if the costs paid by each developer are negotiated separately. Therefore, in-lieu-of charges should not change over time or reflect factors other than user runoff contribution.

Developers may resist these assessments if they can identify improvements that are less expensive than their proportionate share of the planned regional facilities. Administrative costs for this option may also be high.

SYSTEM DEVELOPMENT CHARGES. System development charges (or impact fees) are fees or charges collected from new developments to recover the cost of increasing the capacity of a stormwater or other system to meet the needs of new customer growth. These fees frequently are used to repay debt service on capital improvements and thus could be considered an annual funding option. System development charges are discussed as a financing option because their primary purpose is to generate funds for capital improvements that provide the capacity to serve new development.

Stormwater impact fees are designed to reflect the stormwater contribution of new development and the costs incurred to provide enough additional control facilities in the system to meet those needs. These charges are set in a fee schedule uniformly applied to all new development and are often based on impervious area. In this way, all new development contributes on a consistent basis to the costs of the regional stormwater system rather than to specific regional facilities to serve the individual development. This provides an equitable way, from a regional viewpoint, to recover these costs.

System development charges shift the burden of payment to new residents for facilities sized to serve future growth, rather than imposing these costs on existing customers. These revenues may be used to reimburse existing customers for costs they have incurred to provide capacity for future users. Revenues from these fees may be used to retire a debt issued to finance regional facilities or as part of a pay-as-you-go financing plan. Revenues will fluctuate with the amount of new construction.

COMMUNITY DEVELOPMENT BLOCK GRANTS. This is a federal grant program to aid local governments in constructing various types of public improvements that cannot be financed by any other available source. This program is intended to benefit low-income groups or communities. These grants may be difficult to administer because they originate and are supervised by a federal agency—the Department of Housing and Urban Development. Funds available are also limited.

INTERNAL BORROWING. Another source of revenue for financing the construction or purchase of capital improvements involves borrowing money from other local departments or utilities. The borrowed money must be repaid with interest.

Implementation Process. The evaluation and selection of appropriate revenue alternatives available to implement a stormwater management master plan typically can be accomplished by the following steps:

- Define level of service goals. Categories of problems should be identified and specific performance criteria defined to address the problems.
- Identify program components. Watershed evaluations should be performed to identify and prioritize specific problem areas, including the development of program facility components or activities to comply with the level of service goals.
- Establish program costs. Based on the program components, a schedule of funding and financing requirements can be prepared.
- Identify revenue alternatives. Available revenue alternatives, as discussed in this section, should be reviewed to identify existing and potential sources.
- Develop funding and financing plans. Combine results of the program cost and revenue alternative evaluations to develop specific funding and financing plans.

The process to accomplish these tasks can require several iterations to match level-of-service goals with achievable funding and financing plans.

EVALUATING EFFECTIVENESS

Chapter 3 of this publication provides detailed guidance on assessment, monitoring, and modeling approaches to evaluating program effectiveness. Assessing the program effectiveness means demonstrating that the BMPs adequately protect beneficial uses of the receiving waters. Doing this correctly can be complex, and specific procedures need to be defined for assessment. The assessment typically consists of the following three methods:

- Ecological monitoring evaluates the habitat in the receiving water, determines stresses on the habitat, and defines habitat that reasonably can be supported in the receiving water.
- Water quality monitoring can take place in either the receiving water where the stormwater discharges or in the stormwater discharges before the receiving water.
- Nonconventional monitoring is the enumeration of some quantity other than direct water quality data to infer pollution reduction or water quality improvement. There are many different kinds of indirect measures that can be devised to estimate the success of a particular BMP program. For example, public surveys may show increases in environ-

mental awareness. Another example would be monitoring the amount of used oil collected.

In either case, it is important to establish the specific objective of the monitoring program. There are primarily three distinctly different objectives to consider:

- Characterize the stormwater discharge. Here, the monitoring program basically is set up to provide water quality data for comparison with other databases (for example, to determine if a pollutant warrants special attention in the stormwater management plan).
- Characterize the receiving water during and after stormwater discharges. A monitoring program carried out under this objective could include "biological surveys" and water quality sampling.
- Assess the effectiveness of the stormwater management plan in reducing pollutants. To accomplish this objective, a carefully laid out program is required. Because of the wide variability in stormwater quality, this monitoring program should be complemented with a targeted BMP. Pilot studies provide an excellent opportunity for evaluating targeted BMPs in a well-defined area or watershed.

REFERENCES

American Society of Civil Engineers and Water Environment Federation (1992) *Design and Construction of Urban Stormwater Management Systems.* ASCE Manuals and Reports of Engineering Practice No. 77, New York, N.Y.; WEF Manual of Practice No. FD-20, Alexandria, Va.

Baltimore County Department of Environmental Protection and Resource Management (1989) Regulations for the Protection of Water Quality, Streams, Wetlands and Floodplains. Towson, Md.

Cummins, J. (1988) Maryland Anacostia Basin Fisheries Study. Phase I Interstate Commission on the Potomac River Basin, Rockville, Md.

Galli, F.J. (1989) Peat Sand Filters: A Proposed Stormwater Management Practice for Urbanized Areas. Dep. Environ. Programs, Metropolitan Wash. Council Gov.

Herson, L. (1989) The State of the Anacostia: 1988 Status Report. Metropolitan Washington Council of Governments, Washington, D.C.

Schueler, T.R. (1993) Mitigating the Impacts of Urbanization. In *Implementation of Water Pollution Control Measures in Ontario.* W.J. Snodgrass and J.C. P'ng (Eds.), Univ. Toronto Press Pub., Hamilton, Ont., Can.

Schueler, T.R., and Helfrich, M. (1988) Design of Extended Detention Wet Pond Systems. In *Design of Urban Runoff Controls.* L. Roesner and B. Urbonas (Eds.), Am. Soc. Civ. Eng., New York, N.Y.

U.S. Environmental Protection Agency (1991) *Proposed Guidance Specifying Management Issues for Sources of NPS Pollution in Coastal Areas. Fed. Regist.*, 56 FR 11, 618.

Urban Drainage and Flood Control District (1992) *Urban Storm Drainage Criteria Manual. Volume 3—Best Management Practices, Stormwater Quality*. Denver, Colo.

Yaro, R.D., *et al*. (1988) *Dealing with Change in the Connecticut River Valley: A Design Manual for Conservation and Development*. Lincoln Inst. Land Policy.

Chapter 3 Monitoring, Modeling, and Performance Auditing

The management of urban stormwater quality requires accurate and detailed information about sources and possible effects of stormwater-related pollution. Monitoring, modeling, and performance auditing activities typically are part of a program that is structured to address specific issues, such as permit compliance, pollutant source identification and removal efficiency, or impact assessment. Therefore, a stormwater management program requires coordinated and detailed monitoring procedures, which include planning, data collection, data analysis, and result interpretation and use. Modeling can supplement and, in some cases, replace monitoring efforts with simulations that allow prediction of both discharge and receiving water quality. Auditing performance is a specific application of monitoring (and, in some cases, modeling) that demands standardization of the physical, chemical, climatic, geological, biological, and meteorological parameters reported as part of the audit. This chapter reviews methods for data collection, analysis, and interpretation in monitoring programs; the selection and use of models; and information on parameters that are essential in assessing the performance of best management practices (BMPs).

MONITORING

Monitoring has been an essential element of management programs. Since the adoption of Public Law 92-500 in 1972, holders of National Pollutant Discharge Elimination System (NPDES) permits have been responsible for self-monitoring. Self-monitoring must meet specific frequency and quality criteria, and contain information about discharge quality, toxicity, and receiving water conditions. In 1991, the U.S. Environmental Protection Agency (U.S. EPA) adopted an NPDES permitting program for urban runoff, and in late 1992, U.S. EPA published a draft policy statement on combined sewer overflows. Stormwater NPDES permit monitoring requirements have included the development of sampling programs and the assessment of both discharges receiving water conditions. Further, U.S. EPA has placed a high priority on controlling nonpoint source pollution through clarification of the requirements of Section 303(d) of the Clean Water Act. Dischargers are now faced with total maximum daily load analyses within a water-quality-based toxics control program (U.S. EPA, 1991).

The result of these developments in regulatory programs is an increasing emphasis on monitoring physical, chemical, and biological conditions of receiving water. Holders of NPDES permits, environmental quality managers, consultants, and others must now deal with a regulatory environment that requires monitoring beyond the analysis of chemical water quality parameters in effluents. Different data collection procedures and integrated monitoring efforts are now required to meet the typical objectives of management programs.

This section of the publication has been developed to provide the reader with a summary of the steps that lead to effective and efficient monitoring programs. The design process starts with developing a clear statement of objectives, followed by identifying data needs, data collection procedures, and methods of data analysis. Further, to develop a context for monitoring stormwater, it is necessary to connect storm events with stormwater runoff and recognize that the complex interaction of physics and chemistry on the land as well as in the channel will have an equally complex effect on receiving waters. In short, a watershed approach is essential in stormwater monitoring.

FUNDAMENTAL DESIGN ISSUES. **Study Design Assurance**. Monitoring efforts in stormwater management programs typically are initiated based on a general need to identify sources, quantities, and effects of pollutants. Unfortunately, this need is often expressed as a general direction for studies, not sufficiently detailed objective statements that will direct monitoring efforts. For example, a general statement might specify "assess water quality." That assessment could include water quality sampling and analysis, toxicity testing, or biosurveys. The toxicity testing might include laboratory or *in situ* testing. The biosurvey could sample algae, macroinvertebrates, or fish, including single samples or long-term assessments. Any of these efforts will require different personnel and have different costs. Further, water quality analyses, toxicity testing, or biosurveys may answer different questions about receiving waters and support management decision making in a different way. The first step in a monitoring program will be the careful development of a statement of objectives, which will direct specific data collection efforts.

A monitoring design approach proposed by Schaeffer *et al.* (1984) and Herricks and Schaeffer (1987), termed study design assurance (SDA), provides a workable process to ensure that the monitoring conducted fully meets an agreed objective. The SDA process recognizes that the first step in monitoring program implementation is the review of stated study objectives. If an objective is general (such as the determination of environmental impact), then specific monitoring program objectives (such as the determination of the effects of fish, macroinvertebrates, and water supplies) must be developed to guide data collection. An effective, specific objective is clear and testable, similar to hypotheses that are used to test a general theory in many scientific experiments. This development of specific objectives is an iterative process where the initial objective is used to specify the data needed to meet the objective. Then, the data needs are evaluated and a "hypothetical" database is developed. At this stage, it is possible to determine if sufficient resources (both fiscal and scientific/technical) are available to collect the data needed to meet the stated objective. If resources are limited, the initial general objective is then modified. This modification will produce a more specific objective that reflects the limits in available resources while still addressing the stated objective. A hypothetical database is again identified and evaluated against available resources. At this time, it may be apparent to both the managers and the scientists and technicians that initial management expectations are unrealistic, and management expectations are then modified with the stated objective. The process of objective development, evaluation, and more specific objective development may require many iterations.

Critical to the SDA process is the effective communication of monitoring "realities" to managers and the corresponding resource "limits" to monitoring specialists. Therefore, in the SDA process, managers come to recognize that monitoring realities may constrain initial expectations, leading to a more realistic understanding of how monitoring data can be used in their decision making. The scientific and technical staff benefit from the SDA process because they clearly identify how the data can be used before costly data collection efforts are initiated. Matching expectations to what actually can be achieved with a specific data collection program avoids a typical problem where data limitations fail to meet management needs, and potentially useful information from a monitoring effort is discredited.

The organization of the SDA process calls for iterative analysis at all stages in the management process, formalizing data collection to meet specific decision requirements. Adoption of SDA procedures can ensure that a rigor similar to that imposed by the scientific method in experimentation is applied to monitoring.

Monitoring Program Implementation. Implementation of a monitoring program can be considered a five-step process that includes identification and acquisition of existing data, design of the assessment program, collection of the data, analysis of the data, and presentation of the data/information developed.

IDENTIFY AND ACQUIRE EXISTING DATA. The first step of any monitoring design is to identify and review available data. Obtaining access to existing data will have a number of benefits. First, it is relatively inexpensive; when involved in the development of a monitoring program, fiscal issues will often be the most significant constraint. It may even be possible to use existing data to meet management objectives, and scarce resources can be applied to other needs in the management program. Second, existing data will guide the monitoring design by providing critical information on existing conditions. It will be possible to evaluate "watershed" or other influences on the monitoring site, identify problem areas or areas of significant interest, and, as part of the SDA process, use existing data to better define specific objectives and expectations for monitoring data use. Finally, existing data provide a basis for early comparison of results, supporting mid-course corrections typical in any data collection effort.

PREPARE THE MONITORING DESIGN. Monitoring programs must emphasize information development, not simply data acquisition. Further, the SDA process has, at its foundation, the analysis of cost versus information. Thus, the monitoring program design must clearly identify sampling locations, sampling variables, and frequency of sampling to identify not only the information to be acquired but also the cost of that information. The typical sampling design will identify a reference location, a site in the area of maximum effect, and a location where effects have moderated and the receiving waters have recovered. Sampling variables should include sufficient water quality data to characterize water quality and assist in the interpretation of the results of biological sampling. Biological variables should be selected on the basis of expected effect—benthic macroinvertebrates are often used because they are not mobile, while fish may provide information connected directly to public interests and existing (fisheries) management programs. Sampling frequency will vary, but most bioassessment programs will require seasonal sampling. Finally, added to these design elements are the quality assurance/quality control (QA/QC) program requirements, which include both field and laboratory elements.

COLLECT THE DATA. Although data collection is often identified as one of the primary functions of monitoring design, if the existing data are identified and acquired, needed data may be acquired from existing sources rather than requiring new data collection efforts. The final sampling program design should recognize that any data collection effort in natural systems must deal with conditions that are constantly changing. The data collection program should be flexible enough to allow modifications as these receiving system changes occur. A common problem in many data collections programs should be avoided. In many programs, data collection is often assigned the most junior, lowest-cost, and least-experienced member of the team. Because any analysis will only be as sound as the data on which it is based, good analysis will depend not only on a good QA/QC program, but also on competent staff. Elements of a good data collection program are

- Assignment of competent staff,
- Maintenance of design flexibility,
- Rapid processing of data so that results can be used for continuous improvement in design, and
- Maintenance of a QA/QC program.

ANALYZE THE DATA. Data analysis is a process that requires both specific skills and the benefit of experience. Identifying small anomalies in a data set, developing connections between disparate types of data, and extracting the maximum amount of information from a data set are skills that grow with practice but may develop into standard analysis procedures for longer term monitoring efforts. As mentioned in the discussion of data collection, the analysis of data will often depend on and be constrained by the quality of the collection effort. The process of data analysis includes

- Evaluation of the QA/QC program and its results;
- Analysis of independent factors (parameter-specific trends and identified problems with the data set);
- Analysis of the relationships among independent factors (this analysis typically will involve arrangement of data sets, such as ranking and correlation—it is extremely important to maintain flexibility in this analysis step and not get locked into a limited set of analysis procedures that may not be appropriate for the data collected);
- Comparison of new data with historical data or other data collected as a part of the current monitoring program;
- Creation of an archive; and
- Maintenance of the established QA/QC program.

PRESENT THE DATA. The actual presentation of the data is possibly the most important step in monitoring program implementation because design, collection, and analysis quality are all represented in this program step. In the presentation, one should

- Present the facts simply and in an understandable format;
- Meet the objectives and goals initially set for the monitoring program;
- Translate monitoring complexity (including elements of analysis) without losing meaning; and
- Provide a basis for future use, specifically relating the presentation to the data archive.

Quality Assurance and Quality Control. In the monitoring program implementation steps presented, one of the consistent elements of monitoring activities is the development and implementation of the QA/QC program. The QA/QC program is initiated with the application of the study design process and followed by specific quality control activities suited to data collection, analysis, or interpretation. Quality control procedures can be applied to any data collection effort or monitoring program, but it must be recognized that studies with different objectives involve different approaches to data collection and will have different quality control elements. What should be sought is an internal statistical quality control and an external quality control, proficiency testing, or laboratory evaluation. Both product quality and production processes must be evaluated.

In programs designed to assess receiving system effects of urban stormwater, it will be necessary to consider QA/QC elements specific for biological monitoring. Biological monitoring can be separated into two types of analysis: toxicity testing and biosurveys, or bioassessments. Toxicity tests are laboratory-based analyses that incorporate rigorous experimental protocols operating under strict environmental control to expose selected organisms to toxicants for a defined time period. Guidance for QA/QC programs is provided in *Standard Methods for the Examination of Water and Wastewater* (APHA *et al.*, 1995) and ASTM publications. Bioassessments are field-based analyses that lack strict experimental controls and may range from the description of organisms present in a community or ecosystem to the measurement of a range of ecosystem properties and processes. Bioassessments may include the use of experimental manipulation of contaminants, habitat, and ecological relationships in ecosystems, but bioassessments depend on uncontrolled reference areas to assess the consequence of the manipulation. The QA/QC procedures for biosurveys and bioassessments begin with adherence to quality assurance procedures for field and laboratory analyses of chemical and biological data (U.S. EPA, 1973a, and USGS, 1993). Quality assurance procedures for database development and management, including entry error checking, range checking, and statistical/graphical detection of outliers, should follow current practices for research databases (Gurtz, 1986).

Physical, Chemical, and Biological/Ecological Measurements

Because the actual selection of measurements needed in a monitoring program is objective and often site specific, a full discussion of measurements useful in the assessment of urban runoff effects is not possible in this publication. Nonetheless, it is possible to reference a number of government publications that provide detailed information on measurements typically made in monitoring programs. Publications listed in the Reference and Suggested Readings sections provide specific guidance on measurement and assist in setting the context for analysis and interpretation.

PHYSICAL PARAMETER SELECTION. Changes in physical conditions because of urban runoff include both change in hydrology and modification of channels in the receiving systems. Typical hydrological data measured in monitoring programs include analysis of discharge volume and distribution. From these data, it is possible to develop information about storm runoff peaks, runoff volumes, or storms and base flow. General hydrologic parameters that should be measured or calculated from discharge information include

- Runoff volume parameters during the monitoring season:

  V_R = volume of the average runoff event in the watershed, mm (in.);
  V_{50} = volume of the 50th percentile runoff event in the watershed, mm (in.);
  CV_{VR} = coefficient of variation in the volumes of runoff events ($V_{SD\text{-}R}/V_R$), in which $V_{SD\text{-}R}$ = standard deviation of runoff volumes;
  V_B = volume of the seasonal dry weather base flow in the watershed, mm (in.);
  Q_P = average runoff peak rate, m^3/s (cu ft/sec); and
  CV_{QP} = coefficient of variation of flow peaks.

- Time-variable parameters of storms during the monitoring season.
- Storm runoff interevent (separation) time:

  T_S = average separation period between the end of a storm runoff hydrograph and the beginning of the next one, hours;
  T_{S50} = the 50th percentile of storm runoff event separation periods, hours; and
  CV_{TS} = coefficient of variation in storm runoff event separation periods ($T_{SD\text{-}S}/T_S$), in which $T_{SD\text{-}S}$ = standard deviation of storm runoff event separation periods.

- Storm runoff duration:

  T_D = average duration of storm runoff, hours;
  T_{D50} = the 50th percentile value of storm runoff duration, hours; and
  CV_{TD} = coefficient of variation in storm runoff duration ($T_{SD\text{-}D}/T_D$), in which $T_{SD\text{-}D}$ = standard deviation of storm runoff duration.

The modification of channels should focus on analysis of

- Stream channel widening and downcutting;
- Changes in stream bank erosion;
- Changes in channel features, including bar location and sediment size composition;
- Changes in pool/riffle characteristics; and
- Stream relocation/enclosure or channelization.

Unlike hydrology, where a long history of practice has produced well-defined measurement and analysis methods, the process of monitoring changes in channel morphology is evolving. Accordingly, rather than make definitive statements in this publication, two types of data collection are recommended—indicators of change and establishment of data on stream channel stability (Rhoads, 1995).

Key parameters for change in channel morphology include

- Meander geometry,
- Width:depth ratio, and
- Frequency of full-bank flow.

Systems for classifying the morphology of streams are less well defined. One system, the Rosgen system, is receiving wide application. Other approaches are energy based (unit stream power) or ecosystem based (habitat suitability index modeling). The basic data needed for classification include

- Watershed position (stream order),
- Riparian zone condition (for example vegetation or encroachment),
- Watershed and local geology and land use,
- Channel morphology,
- Channel profile, and
- Substrate and sediment characteristics.

WATER QUALITY PARAMETER SELECTION. In urban runoff, two aspects of water quality are often the focus of monitoring programs. The first is the concentration of contaminants in the runoff, particularly the changes in concentration that occur through time. The second is the eventual loading to the receiving system, which integrates concentration and discharge flow. The parameters selected for monitoring include general parameters such as temperature, alkalinity, hardness, conductivity, and pH and site-specific parameters that will include analyses for conservative and nonconservative pollutants. The actual selection of parameters should be based on expected sources of contamination or should be guided by water quality standards, which are the foundation for any water quality regulation. Of particular importance are contaminants that may bioaccumulate.

Because the presence and concentration of contaminants change rapidly in urban runoff, a critical element of water quality analyses is sampling. For selected general parameters (temperature, dissolved oxygen, pH, and conductivity), continuous monitoring using multiparameter "sonde" units is recommended. Measurement frequency of these general background parameters should be selected based on expected storm hydrograph characteristics, and the duration of the monitoring should reflect pre- and poststorm conditions. Sampling for specific parameters should be accomplished by grab sampling. Automatic samplers are available that can collect sequential grab samples during an event; mix sequential grab samples to provide an event composite sample; or provide flow-weighted or time-weighted single, or multiple, sample composites. Flow-weighted composite sampling is essential for contaminant-loading determinations.

Analysis of water quality parameters should follow standard methods, and QA/QC programs should be in place for both sample collection and sample analysis. Two interpretation issues typically associated with urban runoff analysis are detection limits and outliers. Analyses below detection limits should not be reported as zero values. General guidance suggests that water quality data should not be censored by detection limits; whenever possible, report actual concentration (positive or negative) regardless of whether it is below detection limits (Gilbert, 1987). Care must also be exercised in dealing with outliers. Although excursions well beyond other analytical results are often attributed to analytical error, the nature of urban runoff sampling suggests that these outliers may be "real" measures of short-term events and should be preserved in reported data.

BIOLOGICAL MONITORING. Although guidance for biological monitoring is provided in numerous published sources, the most typically applied procedures are rapid bioassessment protocols (RBPs), which are directed to macroinvertebrate and fisheries sampling. The following discussion of RBP procedures provides guidance for biological monitoring of urban runoff.

RAPID BIOASSESSMENT PROTOCOLS. In 1989, U.S. EPA published a set of protocols for bioassessments in *Rapid Bioassessment Protocols for Use in Streams and Rivers—Benthic Macroinvertebrates and Fish* (U.S. EPA, 1989). The implementation framework first describes the development of an empirical relationship between habitat quality and biological condition. As additional information is obtained from systematic monitoring, a relationship between habitat and biological potential is developed and the effect of water quality alteration can be objectively determined from either habitat change or measures of biological integrity.

Five rapid bioassessment protocols have been developed, three for benthic invertebrates and two for fish. The appropriate bioassessment approach depends on the study objectives. Rapid bioassessment protocols I and IV are screening tools to help determine if biological impairment exists. Benthic RBP I and fish RBP V are more rigorous and provide more objective and reproducible evaluations than RBPs I and IV. Rapid bioassessment protocols II, III, and V are semiquantitative and use an integrated analysis technique to provide continuity in evaluation impairment among sites and seasons. Each of the RBPs is summarized briefly.

- Rapid bioassessment protocol I—benthic macroinvertebrates, and rapid bioassessment protocol IV—fish. These RBPs provide a screening mechanism to identify biological impairment. They are neither intended to quantify the degree of impairment nor provide definitive data to establish cause-and-effect relationships. They allow a cursory assessment, using cost and time efficiencies to evaluate a large number of sites, identify major water quality problems, and help plan and develop management strategies.

- Rapid bioassessment protocol II—benthic macroinvertebrates. This RBP provides information to rank sites as severely or moderately impaired so that additional study or regulatory/management action can be planned. Like RBP I, this protocol can be used as a screening tool and allows agencies to evaluate a large number of sites with relatively little time and effort. The more documented procedures and integrated metrics of RBP II promote better consistency and allow better comparison among sites.
- Rapid bioassessment protocol III—benthic macroinvertebrates, and rapid bioassessment protocol V—fish. These two RBPs provide a consistent, well-documented biological assessment. Like RBP II, they provide information for ranking site impairment and a way to compare repeatable results over time (trend monitoring). These RBPs include taxonomic identifications to the lowest practical level, thereby providing information on population as well as community-level effects. They include an integrated assessment of metrics and can be used to develop biocriteria.

AN INTEGRATED ASSESSMENT APPROACH. An integrated assessment of urban runoff effects should include a watershed-based analysis that connects physical, chemical, and biological/ecological analysis activities in an effective, well-focused monitoring program. An integrated effort initially may focus on water quality, toxicity, physical dynamics, or general system health as program elements, but these analyses eventually must be brought together in a watershed context. Because spatial relationships are critical in an integrated approach to understanding and management of water resource problems caused by urban runoff and nonpoint sources, integrated management requires careful consideration of the spatial framework for the monitoring program. Traditionally, we have relied on spatial frameworks based on political boundaries, watersheds, hydrologic units, or physiographic regions. However, these areas do not correspond to patterns in vegetation, soils, land surface form, land use, climate, rainfall, or other characteristics that control or reflect spatial variations in surface water quality or aquatic organisms. An alternate approach is based on ecoregions. Omernik (1987) proposed using spatial frameworks based on ecological regions (ecoregions) to assess the health of aquatic systems. Ecoregions reflect similarities in the type, quality, and quantity of water resources and the factors affecting them. Therefore, regional patterns of environmental factors reflect regional patterns in surface water quality.

MODELING

URBAN MODELING OBJECTIVES AND CONSIDERATIONS. Studies and projects involving urban stormwater runoff quality can relate to many problems. In the broadest sense, water quality studies may be performed to protect the environment under various state and federal legislation. In a narrower sense, a study may address a particular water quality issue in a particular receiving water, such as the bacterial contamination of a beach or the release of oxygen-demanding material to a stream or river. Many of these studies can be supported or completed through modeling. By no means should it be assumed that every water quality problem requires a water quality modeling effort. Some problems may be mostly hydraulic in nature, for example, problems with basement flooding. That is, the solution may often reside primarily in a hydrologic or hydraulic analysis in which the concentration or load of pollutants is irrelevant. In some instances, local or state regulations may prescribe a nominal "solution" without recourse to water quality analysis. For example, stormwater runoff in the state of Florida is considered "controlled" through retention or detention with filtration of the runoff from the first inch of rainfall for areas of 40 ha (100 ac) or less. Other problems may be resolved through the use of measured data without the need to model. In other words, many problems do not require water quality modeling at all.

If a problem does require modeling, some objectives are better met through modeling than others. Models may be used for objectives such as the following:

- Characterize the urban runoff as to temporal and spatial detail, and concentration/load ranges;
- Provide input to a receiving water quality analysis (for example, drive a receiving water quality model);
- Determine effects, magnitudes, locations, and combinations of control options;
- Perform frequency analysis on quality parameters (to determine return periods of concentrations/loads); and
- Provide input to cost–benefit analyses.

The first two objectives characterize the magnitude of the problem and the last four objectives are related to the analysis and solution of the problem. Computer models allow some types of analysis, such as frequency analysis, to be performed that could rarely be performed because the history of water quality measurements in urban areas is often poor. It should always be recognized, however, that the use of measured data is typically preferable to the use of simulated data, particularly for the first two objectives, in which accurate concentration values are needed. Typically, models are not good substitutes for good field-sampling programs. On the other hand, models can sometimes be used to extend and extrapolate measured data and enhance field-sampling results.

Careful consideration should be given to providing input to a receiving water quality analysis. The model output needed to drive a receiving water quality model is related to the objective of the analysis program. Modeling can also follow a modified SDA process. If the focus in a monitoring effort is toxicity, the model should produce concentration versus time predictions for short intervals (5 to 15 minutes). In fact, the first urban runoff quality model incorporated the concept of simulation of detailed intrastorm quality variations, for example, the production of a "pollutograph" (concentration versus time) at 5- or 10-minute intervals during a storm for input to a receiving water quality model. If the objective of the receiving water quality modeling effort is to assess general effect, the modeling effort should be supported by a bioassessment program and the model need not predict short-term variations in concentration. The total storm load will be sufficient to determine the receiving water response.

Differences in detail require differences in model complexity, as seen in Table 3.1. The most complex models are needed to predict concentration versus time at high frequencies. If only the total storm loads are needed, this presents a much easier modeling task.

In any modeling effort, data requirements are critical. Such requirements may be as simple as a constant concentration, or they may include detailed time-related changes in concentration or flow. Data may be obtained from existing studies or require extensive field monitoring. For some model objectives, it may not be possible to actually measure fundamental input parameters, which are obtained through model calibration. Acquisition of the high-quality data needed to support modeling efforts, either through literature reviews or field surveys, will affect the level of effort and costs associated with the management program. Details on data requirements for model appli-

Table 3.1 Required temporal detail for receiving water analysis.

Type of receiving water	Key constituents	Frequency
Lakes, bays	Nutrients and toxics	Weeks, seasonal, years
Estuaries	Nutrients, oxygen demand, bacteria, toxics	Tidal cycle, days, weeks
Large rivers	Oxygen demand, nitrogen, toxics	Minutes, days (event based)
Streams	Oxygen demand, nitrogen, bacteria, toxics	Minutes, hours, days (event based)
Ponds	Oxygen demand, nutrients, toxics	Hours, weeks
Beaches	Bacteria	Hours (event based)

cation in urban areas will be deferred until modeling techniques are described.

OVERVIEW OF AVAILABLE MODELING OPTIONS. Several quality modeling options exist for the simulation of quality in urban storm and combined sewer systems. These have been reviewed by Huber (1985 and 1986) and range from simple to involved, although some "simple" methods (such as U.S. EPA statistical methods) can incorporate quite sophisticated concepts. The principal methods available to the contemporary engineer are outlined generically below, in a rough order of complexity. Their data requirements are summarized again in a following section.

- Constant concentration or unit loads,
- Spreadsheet,
- Statistical,
- Rating curve or regression, and
- Buildup/washoff.

Constant Concentration or Unit Loads. As its name implies, constant concentration means that all runoff is assumed to have the same constant concentration at all times for a given pollutant. At its simplest, an annual runoff volume can be multiplied by a concentration to produce an annual runoff load. However, this option may be coupled with a hydrologic model, wherein loads (product of concentration and flow) will vary if the model produces variable flows. This option may be useful because it may be used with any hydrologic or hydraulic model to produce loads simply by multiplying by the constant concentration. In many instances, it may be most important to get the volume and timing of such overflows and diversions correctly and simply estimate loads by multiplying by a concentration.

An obvious question is which constant concentration to use. Early (pre-1977) concentration and other data are summarized in publications such as Lager *et al.* (1977) and Manning *et al.* (1977). U.S. EPA Nationwide Urban Runoff Program (NURP) studies (U.S. EPA, 1983) have produced a large and invaluable database from which to select numbers, but the 30-city coverage of NURP will most often not include a site representative of the area under study. Nonetheless, a large database does exist from which to review concentrations. Another option is to use measured values from the study area. This might be done from a limited sampling program.

Unit loads are perhaps an even simpler concept. These consist of values of mass per area per time (typically pounds per acre per year or kilograms per hectare per year) for various pollutants, although other normalizations such as pound/curb-mile are sometimes encountered. Annual (or other time unit) loads are thus produced by multiplying mass per time by the contributing area. Such loadings are site specific and depend on both demographic and hydrologic factors. A unit load must be based on an average or "typical" runoff volume and cannot vary from year to year, but loading can be subject to reduction by BMPs if the BMP effect is known. Although early U.S. EPA references provide some information for various land uses (McElroy *et al.*, 1976; U.S. EPA, 1973b; and U.S. EPA, 1976a), unit loading rates are variable and difficult to transpose from one area to another. Constant concentrations can sometimes be used for this purpose because mg/L × 0.226 5 = lb/ac per inch of runoff. Thus, if a concentration estimate is available, the annual loading rate may be calculated by multiplying by the inches per year of runoff.

The universal soil loss equation (Heaney *et al.,* 1975, and Wischmeier and Smith, 1958) was developed to estimate tons per acre (kilograms per square metre) per year of sediment loss from land surfaces. If a pollutant may be considered as a fraction ("potency factor") of suspended solids concentration or load, this offers another option for the prediction of annual loads. Lager *et al.* (1977), Manning *et al.* (1977), and Zison (1980) provide summaries of such values.

Spreadsheets. Microcomputer spreadsheet software is now ubiquitous in engineering practice. Extensive and sophisticated engineering analysis is routinely implemented on spreadsheets, and water quality simulation is no exception. The spreadsheet may be used to automate and extend the constant concentration or unit load determinations. In the typical application of the spreadsheet approach, runoff volumes are calculated simply, typically by multiplying runoff coefficient times rainfall depth. The coefficient may vary according to land use, but the hydrology is inherently simplistic in the spreadsheet predictions. The runoff volume is then multiplied by a constant concentration to predict runoff loads. Alternatively, unit loads are input directly and then multiplied by corresponding land-use areas. The spreadsheet approach is best suited to estimation of long-term loads, such as annual or seasonal, because simple prediction methods typically perform better over a long averaging time and poorly at the level of a single storm event.

The advantage of the spreadsheet is that a mixture of land uses (with varying concentrations or loads) may easily be simulated, and an overall load and flow-weighted concentration can be estimated for the study area (Walker *et al.*, 1989). A study area may range from a single catchment to an entire urban area, and delivery ratios can be added to simulate loss of pollutants along drainage pathways between the simulated land use and receiving waters. The relative contributions of different land uses may be easily identified, and handy spreadsheet graphics tools may be used for displaying the results.

As an enhancement, control options may be simulated by application of a constant removal fraction for an assumed BMP. Although spreadsheet computations can be complex, BMP simulation is rarely more complicated than a simple removal fraction because anything further would require simulation of the dynamics of the removal device (for example, a wet detention pond), which typically is beyond the scope of the hydrologic component of the spreadsheet model. Nonetheless, the spreadsheet can be used to estimate the effectiveness of control options. Loads with and without controls can be estimated, and problem areas can be determined by separate analysis of contributing basin and land-use characteristics. Because most engineers are familiar with spreadsheets, such models can be developed in house.

Again, the question arises of what concentrations or unit loads to use, this time potentially for multiple land uses and subareas. Again, the NURP database will typically be the first one to turn to, with the possibility of local monitoring to augment it.

Statistical Method. The so-called "U.S. EPA Statistical Method" is somewhat generic and until recently was not implemented in any off-the-shelf model or implemented well in any single report (Hydroscience, 1979, and U.S. EPA, 1983). A FHWA study (Driscoll *et al.,* 1989) partially remedies this situation. The concept is straightforward, namely that of a derived frequency distribution for estimated mean concentration (EMC). This idea has been used extensively for urban runoff quantity (Howard, 1976; Loganathan and Delleur, 1984; and Zukovs *et al.*, 1986) but not as much for quality predictions.

The U.S. EPA statistical method uses the fact that EMCs are not constant but tend to exhibit a log-normal frequency distribution. When coupled with an assumed distribution of runoff volumes (also log-normal), the distribution of runoff loads may be derived. When coupled again to the distribution of stream flow, an approximate (log-normal) probability distribution of instream concentrations may be derived (Di Toro, 1984)—a useful result, although assumptions and limitations of the method have been pointed out by Novotny (1985) and Roesner and Dendrou (1985). Further analytical methods have been developed to account for storage and treatment (Di Toro and Small, 1979, and Small and Di Toro, 1979). The method was used as the primary screening tool in U.S. EPA NURP studies (U.S. EPA, 1983) and has also been adapted to combined sewer overflows (Driscoll, 1981) and highway-related runoff (Driscoll *et al.,* 1989). This latter publication offers a concise explanation of the procedure and assumptions and includes spreadsheet software for easy implementation of the method.

A primary assumption is that EMCs are distributed log-normally at a site and across a selection of sites. The concentrations may thus be characterized by their median value and by their coefficient of variation (*CV*—standard deviation divided by the mean). There is little doubt that the log-normality assumption is good (Driscoll, 1986), but similar to the spreadsheet approach, the method typically is then combined with weak hydrologic assumptions (for example, the prediction of runoff using a runoff coefficient). (The accuracy of a runoff coefficient increases as urbanization and imperviousness increase.) However, because many streams of concern in an urban area consist primarily of stormwater runoff during wet weather, the ability to predict the distribution of EMCs is useful for the assessment of levels of exceedance of water quality standards. The effect of BMPs can again be estimated crudely through constant removal fractions that lower the EMC median, but it is harder to determine the effect on the coefficient of variation. Overall, the method has been successfully applied as a screening tool.

Inputs to the method discussed by Driscoll *et al.* (1989) include statistical properties of rainfall (mean and coefficient of variation of storm event depth, duration, intensity, and interevent time), area, and runoff coefficient for the hydrologic component, plus EMC, median, and coefficient of variation for the pollutant. Generalized rainfall statistics have been calculated for many locations in the U.S. Otherwise, the U.S. EPA SYNOP model (Hydroscience, 1979; U.S. EPA, 1976b; U.S. EPA, 1983; and Woodward-Clyde Consultants, 1989) must be run on long-term hourly rainfall records. If receiving water effects are to be evaluated, the mean and *CV* of the streamflow are required plus the upstream concentration. A lake impact analysis is also possible based on phosphorus loadings.

As with the first two methods discussed, the choice of median concentration may be difficult, and the statistical method requires a coefficient of variation as well. Fortunately, from NURP and highway studies, *CV* values for most urban runoff pollutants are fairly consistent, and a value of 0.75 is typical. If local and/or NURP data are not available or inappropriate, local monitoring may be required.

Regression—Rating Curve Approaches. With the completion of the NURP studies in 1983, there are measurements of rainfall, runoff, and water quality at more than 100 sites in more than 30 cities. Some regression analysis has been performed to try to relate loads and EMCs to catchment, demographic, and hydrologic characteristics (Brown, 1984; McElroy *et al.*, 1976; and Miller *et al.*, 1978), the best of which are results of the U.S. Geological Survey (USGS) (Driver and Tasker, 1988, and Tasker and Driver, 1988) and are described briefly in this section. Regression approaches have also been used, with limited success, to estimate dry-weather pollutant deposition in combined sewers (Pisano and Queiroz, 1977). What are termed "rating curves" in this discussion are a special form of regression analysis in which concentration or loads are related to flow rates or volumes.

A rating curve approach is most often applied using total storm event load and runoff volume, although intrastorm variations can sometimes also be simulated (for example, Huber and Dickinson, 1988). It is typically observed (Driscoll *et al.*, 1989; Huber, 1980; and U.S. EPA, 1983) that concentration (EMC) is poorly correlated or not correlated with runoff flow or volume, implying that a constant concentration assumption is adequate. Because the load is the product of concentration and flow, load typically is well correlated with flow regardless of whether concentration correlates well. This instance of spurious correlation (Bensen, 1965) is often ignored in urban runoff studies. If load is proportional to flow to the first power (that is, linear), then the constant concentration assumption holds; if not, some relationship of concentration with flow is implied. Rating curve results can be used by themselves for load and EMC estimates and can be incorporated to some models.

Rainfall, runoff, and quality data were assembled for 98 urban stations in 30 cities (NURP and others) in the U.S. for multiple regression analysis by USGS (Driver and Tasker, 1988, and Tasker and Driver, 1988). Thirty-four multiple-regression models (mostly log-linear) of storm runoff constituent loads and storm runoff volumes were developed, and 31 models of storm runoff EMCs were developed. Regional and seasonal effects were also considered. The two most significant explanatory variables were total storm rainfall and total contributing drainage area. Impervious area, land use, and mean annual climatic characteristics also were significant explanatory variables in some of the models. Models for estimating loads of dissolved solids, total nitrogen, and total ammonia plus organic nitrogen (total Kjeldahl nitrogen) typically were the most accurate, whereas models for suspended solids were the least accurate. The most accurate models were those for the more arid Western U.S., and the least accurate models were those for areas that had large mean annual rainfall.

These USGS equations represent some of the best generalized regression equations available for urban runoff quality prediction. Note that these equations do not require preliminary estimates of EMCs or local quality monitoring data, except of verification of the regression predictions. Regression equations only predict the mean and do not provide the frequency distribution of a predicted variable, a disadvantage compared to the statistical approach. (The USGS documentation describes procedures for calculation of statistical error bounds, however.) Finally, regression approaches, including rating curves, are difficult to apply beyond the original data set from which the relationships were derived. That is, they are subject to potential errors when used to extrapolate to different conditions. Thus, the usual caveats about use of regression relationships continue to hold when applied to prediction of urban runoff quality.

Buildup and Washoff. In the late 1960s, a Chicago study by the American Public Works Association (1969) demonstrated the (assumed linear) buildup of "dust and dirt" and associated pollutants on urban street surfaces. During a similar time frame, Sartor and Boyd (1972) demonstrated buildup mechanisms on the surface and an exponential washoff of pollutants during rainfall events. These concepts were incorporated to an early hydraulic model (Metcalf and Eddy, Inc., *et al.*, 1971) as well as to the other models to a greater or lesser degree (Huber, 1985). "Buildup" is a term that represents all of the complex spectrum of dry-weather processes that occur between storms, including deposition, wind erosion, and street cleaning. The idea is that all such processes lead to an accumulation of solids and other pollutants that are then "washed off" during storm events.

Although ostensibly empirically based, models that include buildup and washoff mechanisms use conceptual algorithms because the fundamental physical foundations are related to principles of sediment transport and erosion that are sometimes poorly understood. Furthermore, the inherent heterogeneity of urban surfaces leads to the use of average buildup and washoff parameters that may vary significantly from conditions in isolated locations such as a street gutter. Thus, except in rare instances where actual measurements of accumulations of surface solids are available, the use of buildup and washoff formulations involve a calibration exercise against measured end-of-pipe quality data. In the absence of accumulation measurements, inaccurate predictions can be expected.

Different models offer different options for conceptual buildup and washoff mechanisms. In fact, with calibration, good agreement can be produced between predicted and measured concentrations and loads with such models, including intrastorm variations that cannot be duplicated with most of the methods discussed earlier. (When a rating curve is used instead of buildup and washoff, it is also possible to simulate intrastorm variations in concentration and load.) A survey of linear buildup rates for many pollutants by Manning *et al.* (1977) is a source of generalized buildup data, and some information is available in the literature to aid in selection of washoff coefficients (Huber, 1985, and Huber and Dickinson, 1988). However, such first estimates may not even get the user in the ballpark (that is, quality—not quantity—predictions may be off by more than an order of magnitude); the only way to be sure is to use local monitoring data for calibration and verification. Thus, as for most of the other quality prediction options discussed in this section, the buildup–washoff model may provide adequate comparisons of control measures or ranking of loads, but it cannot be used for prediction of absolute values of concentrations and loads (for example, to drive a receiving water quality model without adequate calibration and verification data).

It is relatively easy to simulate potential control measures, such as street cleaning and surface infiltration, using this modeling approach. When intrastorm variations in concentration and load must be simulated (as opposed to total storm event EMC or load), buildup and washoff offers the most flexibility. This is sometimes important for the design of storage facilities in which first-flush mechanisms may be influential.

The data for buildup and washoff modeling are sparse (Manning *et al.*, 1977), and the needed measurements are seldom made as part of a routine monitoring program. For buildup, normalized loadings, such as mass/day per area, mass/day per curb-length, or just mass/day, are required, along with an assumed functional form for buildup versus time, such as linear or exponential. For washoff, the relationship of washoff rate (mass/time) runoff rate must be assumed, typically in the form of a power equation. When end-of-pipe concentration and load data are all that are available, all buildup and washoff coefficients end up being calibration parameters.

Related Mechanisms. In the discussion above, washoff rate is assumed proportional to the runoff rate, as for sediment transport, but erosion from pervious areas may be proportional to the rainfall rate. One model includes this mechanism in its algorithms for erosion of sediment from pervious areas. Another includes a weaker algorithm based on the Universal Soil Loss Equation (Heaney *et al.,* 1975, and Wischmeier and Smith, 1958).

Many pollutants, particularly metals and organics, are adsorbed to solid particles and are transported in particulate form. The ability of a model to include "potency factors" or "pollutant fractions" enhances the ability to estimate the concentration or load of one constituent as a fraction of that of another, such as solids (Zison, 1980).

Groundwater contribution to flow in urban areas can be important in areas with unlined and open-channel drainage. The precipitation load may be input to some models, typically as a constant concentration. Point source and dry-weather flow (base flow) loads and concentrations can also be input to some models to simulate background conditions. Other quality sources of potential importance include catch basins and snowmelt.

Scour and deposition within the sewer system can be important in combined sewer systems and some separate storm sewer systems. The state of the art in simulation of such processes is poor (Huber, 1985).

SUMMARY OF DATA NEEDS. In the application of most models, there are two fundamental types of data requirements. First, there are data needed simply to make the model function, that is, input parameters for the model. These typically include rainfall information, area, imperviousness, runoff coefficient, and other quantity prediction parameters, plus quality prediction parameters such as constant concentration, constituent median and *CV,* regression relationships, and buildup and washoff parameters. In other words, each model will have a fundamental list of required input data.

The second type of information is required for calibration and verification of more complex models, namely, sets of measured rainfall, runoff, and quality samples with which to test the model. Such data exist (for example, Driver *et al.*, 1985; Huber *et al.*, 1982; and Noel *et al.*, 1987) but seldom for the site of interest. If the project objectives absolutely require such data (for example, if a model must be calibrated to drive a receiving water quality model), then monitoring may be necessary to produce needed data.

SELECTING URBAN RUNOFF QUALITY MODELS. This summary will relate primarily to quality prediction and will not represent a comprehensive statement of data needs for quantity prediction. However, because rainfall and runoff are required for virtually every study, certain quantity-related parameters are also necessary.

Modeling Fundamentals. Modeling caveats and an introduction to modeling are presented by several authors, including Huber (1985 and 1986), James and Burges (1982), and Kibler (Ed.) (1982), and summarized in *Combined Sewer Overflow Pollution Abatement* (WPCF, 1989). Space does not permit a full presentation here; a few items are highlighted in the following bullets.

- Have a clear statement of project objectives. Verify the need for quality modeling. (Perhaps the objectives can be satisfied without quality modeling.)
- Use the simplest model that will satisfy the project objectives. Often, a screening model, such as regression or statistical, can determine whether more complex simulation models are needed.
- To the extent possible, use a quality prediction method consistent with available data. This would often rule against buildup–washoff formulations, although these might still be useful for detailed simulation, especially if calibration data exist.
- Only predict the quality parameters of interest and only over a suitable time scale. That is, storm event loads and EMCs typically will represent the most detailed prediction requirement, and seasonal or annual loads will sometimes be all that are required. Do not attempt to simulate intrastorm variations in quality unless necessary.
- Perform a sensitivity analysis on the selected model and familiarize yourself with the model characteristics.
- If possible, calibrate and verify the model results. Use one set of data for calibration and another independent set for verification. If no such data exist for the application site, perhaps they exist for a similar catchment nearby.

Operational Models. Implementation of an off-the-shelf model or method will be easiest if the model can be characterized as "operational" in the sense of the following:

- Documentation—this should include a user's manual, explanation of theory and numerical procedures, data needs, and data input format. Documentation most often separates the many computerized procedures found in the literature from a model that can be accessed and easily used by others.
- Support—this is sometimes provided by the model developer but often by a federal agency such as U.S. EPA.
- Experience—every model must be used a "first time," but it is best to rely on a model with a proven track record.

Models described below are operational in this sense. New methods and models are constantly under development and should not be neglected simply because they lack one of these characteristics, but the user should be aware of potential difficulties if any characteristic is lacking.

Surveys of Operational Urban Runoff Models. Several publications, though somewhat out of date, provide reviews of available models. Some models have persisted for many years and are included in both older and newer reviews, while other models are more recent. Reviews that consider surface runoff quality models include Barnwell (1984 and 1987), Bedient and Huber (1988), Huber (1985 and 1986), Huber and Heaney (1982), Kibler (Ed.) (1982), Viessman *et al.* (1989), WPCF (1989), and Whipple *et al.* (1983).

URBAN RUNOFF QUALITY SIMULATION MODELS AND METHODS. Several models are often considered the best choices for full-scale simulation for urban areas. Other models have been adapted and given modified names. Still other models have been used for water quality simulation for a specific project (Noel and Terstriep, 1982), but such modifications and quality procedures remain undocumented, and the quality model cannot be considered operational. A number of models have been developed in Europe and applied extensively in a number of situations. One of these models includes modules for generation of runoff from rainfall using either a time–area method or a nonlinear reservoir model. The model handles six pollutants, and the emphasis is production of statistics for both extreme and annual loads. Finally, there are many models well known in the hydrologic literature that are useful in the hydrologic modeling in water quality studies but do not simulate water quality directly.

BIOLOGICAL AND ECOLOGICAL MODELING. In many respects, biological and ecological modeling have lagged far behind physical, chemical, and hydrologic modeling, and there are few examples of general application of biological and ecological models in urban runoff. It is possible to connect other modeling efforts with a prediction of biological and ecological receiving system effects through a general application of concentration and time of exposure analysis, which allows assessment of toxicity and supports general predictions of effect.

RECEIVING WATER MODELS

To assess the effects of runoff loads on receiving water quality, it is often necessary to use computer models. Measurements of receiving water quality parameters are preferable for impact assessment, but such data typically are sparse. Also it is difficult and costly to obtain sufficient data for model calibration, let alone for evaluation of effects. And if alternative pollution control options are to be evaluated, models are the only option with which to assess "what if" management strategies, at least as far as effects on *in situ* concentrations are concerned. It sometimes may be possible to evaluate water quality control strategies on the basis of hydraulic or surface runoff quality criteria alone, without receiving water quality modeling—an advantage.

Do not model if it is not necessary. However, if comparison with water quality standards is a requirement, modeling typically is the only option.

Receiving water quality models are available for streams, lakes (and reservoirs), estuaries, and bays. Groundwater models will not be discussed here, although several models are available. Segments of coastal ocean areas can also be modeled with more difficulty (because of the necessity of two- and three-dimensional formulations). Such models are driven by transient or steady-state point and nonpoint source loadings, typically entering the water body at multiple locations. Nonpoint source loads are often generated by surface runoff loading models discussed previously. Thus, there typically is a coupling of surface runoff models with receiving water models for determination of nonpoint source effects. The output from such models typically is a transient or steady-state prediction of water quality constituent concentrations at multiple locations throughout the receiving water, although some methods (such as simple eutrophication models) provide only an average concentration in space and time.

Most receiving water quality models require information about flows, velocities, volumes, and stages—that is, a description of quantity (hydrodynamic) processes. Some models compute flows and quality concentrations, whereas other models require a separate model or data for input of such information. Data input for the former is correspondingly more demanding. This discussion will focus only on quality modeling because of the vast complexity of two- and three-dimensional quantity modeling. Simulation of quantity processes will thus be mentioned only incidentally if it is included as an option for a particular model.

Most of the caveats and general modeling fundamentals provided previously while discussing surface runoff models also apply to receiving water quality modeling, for example clear objectives, desirability of simplicity, and the need for data. Receiving water quality models should be calibrated and verified similar to surface runoff models because of the many influential processes at work in natural waters. A similar statement can be made for receiving water quality modeling as for surface runoff quality modeling: calibration/verification data (that is, measured *in situ* concentrations) are essential for accurate predictions of concentrations. Without such data, only relative comparisons can be made. Of course, accurate concentrations are important for comparison of predicted concentrations with water quality standards.

The models discussed below are all "far-field" models, that is, models for which transport is influenced only by the hydrodynamics of the receiving waters. "Near-field" models consider the effects of plumes and jets at the discharge location and may be used for mixing zone calculations and are not considered herein. These dilution calculations are hydrodynamic in nature and discussed by Fischer *et al.* (1979) and Holley and Jirka (1986). Another category of models not included below is simple eutrophication models for lakes. Procedures for analyzing lakes in a spatially lumped manner for eutrophication screening are described, for example by Mills *et al.* (1985), Reckhow and Chapra (1983), and Thomann and Mueller (1987).

LINKAGE WITH SURFACE RUNOFF MODELS. Nonpoint source loadings used to "drive" the receiving water model typically are obtained from surface runoff models. A time series of loads (and flows) must be supplied as input to the receiving water model. This interface may be more or less difficult depending on the models used. The user should determine the nature of the interface requirements to ensure compatibility between the surface and receiving water models. For instance, the surface runoff model should be able to output a file containing the load and flow time series, and it should be possible to manipulate this file to produce the required format for the receiving water model. Obviously, the receiving water model documentation should specify this format. It can thus be seen that the interfacing of different models may be more or less difficult, depending on the models and their documentation. Use of surface and receiving water models from the same agency may alleviate this problem to a large extent.

SURVEY OF RECEIVING WATER QUALITY MODELS. Modeling is such a dynamic process that reviews and surveys are rapidly outdated. Typically, it is best to contact an agency or model distributor directly for current information. However, useful summaries are provided by Ambrose and Barnwell (1989), Ambrose *et al.* (1988), and Feldman (1981). Principles of fate and transport are discussed in references such as Bowie *et al.* (1985), Fischer *et al.* (1979), French and McCutcheon (Eds.) (1989), Holley and Jirka (1986), Krenkel and Novotny (1980), Mills *et al.* (1985), and Thomann and Mueller (1987).

SOURCES OF RECEIVING WATER MODELS. Models are available from federal agencies and private vendors. For water quality formulations (as opposed to detailed hydrodynamic components), the predominant federal source is the U.S. EPA Center for Exposure Assessment Modeling at Athens, Georgia. Additional federal sources include the U.S. Army Corps of Engineers, Waterways Experiment Station in Vicksburg, Mississippi, and the U.S. Army Corps of Engineers, Hydrologic Engineering Center, in Davis, California. The USGS has performed numerous receiving water quality studies, including development of two- and three-dimensional transient hydrodynamic/quality models. Most of these efforts are for a particular project, and local centers should be contacted for information about particular model availability. Similar remarks can be made about the National Oceanic and Atmospheric Administration for estuary/bay modeling and for the Tennessee Valley Authority for river/reservoir modeling. Still other federal agencies may perform quality modeling as a part of other studies.

BEST MANAGEMENT PRACTICE DATA REPORTING AND MONITORING

Auditing and monitoring BMPs are important applications of monitoring and modeling. This section proposes a standardized set of BMP data for reporting purposes and equipment considerations for monitoring specific BMP facilities based on Urbonas (1995a and 1995b).

NEED FOR STANDARDIZED REPORTING. There is a need to develop an approach for reporting data on the physical, chemical, climatic, geological, biological/ecological, and meteorological parameters in the assessment of the performance of BMPs used to enhance urban runoff quality. Transferability of performance results and consistency or lack of it in the performance of various BMPs have been ongoing problems. By defining a standardized approach it is expected that over time this standardization will conserve the resources being expended by various field investigations and will lead to improvements in the selection and design of various BMPs. Further, this standardization provides a workable listing of parameters that are the focus of monitoring and modeling efforts.

FACILITIES COVERED FOR REPORTING. A standard reporting format is provided for the following BMP technologies:

- Retention basin (dry pond, wet pond),
- Extended detention basin,
- Wetland basin,
- Wetland channels,
- Sand filters,
- Oil–grit separators (traps), and
- Infiltration and percolation facilities.

For facilities that contain more than one type of BMP in a BMP train, the analyst should report data for the overall unit, if performance assessment occurs on this basis, and for specific BMPs, if monitoring is done on this basis.

PARAMETERS FOR RETENTION PONDS. Retention ponds always have some surcharge detention storage above the permanent pool water surface. There are several pollutant removal mechanisms at work within a retention pond. These include sedimentation during runoff events and between runoff events and other physical, chemical, and biological processes. As a result, more information needs to be reported for these types of facilities than for facilities that remove pollutants primarily though physical processes. The following parameters emerge as important reporting points to assess retention pond removal efficiency.

Surface area and pond layout parameters:

A_P = surface area of the permanent pool, m^2 (sq ft);
A_L = surface area of the littoral zone (zone ≤ 0.5 m [1.5 ft] deep), m^2 (sq ft);
A_D = surface area of the top of the surcharge detention basin, m^2 (sq ft);
L_P = length of the permanent pool or flow path, m (ft);
L_D = length of the surcharge detention basin, m (ft);
A_F = surface area of the forebay, m^2 (sq ft); and
L_F = length of the forebay, m (ft).

Basin volume parameters:

V_P = volume of the permanent pool, m^3 (cu ft);
V_D = design volume of the surcharge detention basin above the permanent pool's water surface, m^3 (cu ft); and
V_F = volume of the forebay, m^3 (cu ft).

Emptying time parameters:

T_E = time needed to empty 99% of V_D assuming no inflow takes place while the surcharge pool is emptying, hours; and
$T_{0.5E}$ = time needed to empty the upper one-half of V_D assuming no inflow takes place while the surcharge pool is emptying, hours.

PARAMETERS FOR EXTENDED DETENTION BASINS. Extended detention basins use sedimentation as their primary pollutant removal mechanism. As a result, extended detention basins have to be viewed somewhat differently than retention ponds. In a retention pond, sediments that settle below the overflow outlet level are essentially trapped within the permanent pool and are less likely to be discharged through the outlet. The trapped sediment continues to settle to the bottom of the pond even after the surcharge volume is drained off. In an extended detention basin, stormwater empties through an outlet located on the bottom. As the sediments settle to the bottom, they concentrate within the lower levels of the ever-shrinking pool and discharge through the outlet. Unless they are scoured out, only the sediments that deposit on the bottom can be trapped within the basin.

The list that follows reflects most of the parameters of importance for an extended detention basin. Many of the same parameters that were recommended for retention ponds are repeated.

Surface area and plan layout parameters:

A_D = surface area of the extended detention basin, m^2 (sq ft);
L_D = length of the extended detention basin, m (ft);
A_B = surface area of the bottom stage (that is lower basin), m^2 (sq ft); and
L_F = length of the forebay, m (ft)

Basin volume parameters:

V_D = total volume of the extended detention basin, m^3 (cu ft);
V_B = volume of the bottom stage only of the basin, m^3 (cu ft); and
V_F = volume of the forebay, m^3 (cu ft).

Time variables: Use the same emptying time parameters as defined for the retention pond.

PARAMETERS FOR WETLAND BASINS. Some wetland basins are similar in their operation to retention ponds, while others resemble extended detention basins. The difference between the two is whether or not the wetland basin has standing water or a wetland meadow as its bottom. The pollutant removal mechanisms are probably similar to those found in retention ponds and in detention basins, except that stormwater comes in contact with wetland flora and fauna. This contact and the physical structure of the wetland provide pollutant removals through adsorption and biochemical processes and possibly through reoxygenation of the sediments and detoxification of the water column—processes that may or may not be available in retention ponds and that are not available in detention basins.

Each performance monitoring program should report parameters that are particular to the wetland studied. Most currently available wetland monitoring data rarely contain such information, often not even reporting many of the parameters typically reported for other BMPs. Because the quantification of wetland performance as a BMP is relatively new, little information can be found in the literature, and it is difficult to suggest parameters to report when reporting the performance data of wetland basins. Table 3.2 and the follow-

Table 3.2 Additional general parameters to report for wetlands.

Type of wetland	Cattail marsh, northern peat land, meadow, palustrine, southern marshland, hardwood swampland, brackish marsh, high-altitude riverine, freshwater riverine, constructed or natural wetlands
Rock filter?	Is there a rock filter media present in the wetland bottom?
Dominant plant species	Lists the dominant plant species in the wetland and the age of these plants (that is, the time since their original planting or replanting)

ing list suggest the parameters that appear to be most important, many of which are identical to those recommenced for retention ponds.

Surface area and layout plan parameters:

A_P = surface area of permanent wetland pool, if any, m^2 (sq ft);
A_M = surface area of the meadow wetland, if any, m^2 (sq ft);
$P_{0.30}$ = percent of permanent pool less than 0.30 m (12 in.) depth;
$P_{0.60}$ = percent of the permanent pool more than 0.60 m (24 in.) depth;
A_S = surface area of the surcharge detention basin's top, m^2 (sq ft),
L_S = length of the wetland surcharge/detention pool or flow path, m (ft);
A_F = surface area of the forebay, m^2 (sq ft); and
L_F = length of the forebay, m (ft).

Basin volume parameters:

V_P = volume of the permanent pool, if any, m^3 (cu ft);
V_D = design volume of the surcharge/detention basin, m^3 (cu ft); and
V_F = volume of the forebay, m^3 (cu ft).

Time variables: Use the same emptying time parameters as defined for the retention pond.

PARAMETERS FOR WETLAND CHANNELS. Channels can be designed to have a wetland bottom that is designed to flow slowly. When properly designed, the channel's bottom is covered by wetlands, with only the sideslopes having terrestrial vegetation. The flow velocity is controlled by transverse berms, check dams, or an outlet at the downstream end of a given channel's reach. In the last case, the channel is essentially a long and narrow wetland basin. Figure 3.1 shows a profile of an idealized wetland bottom channel.

The pollutant removal mechanisms in wetland bottom channels are similar to those found in wetland basins, except that contact time of stormwater with the wetland vegetation is likely to be less. Because of the flowing channel nature of this BMP, the following parameters should provide some of the information needed to compare the performance of different installations:

$V_{2\text{-yr}}$ = average channel velocity during a 2-year runoff event, m/s (ft/sec);
A_D = surface area of the wetland bottom, m^2 (sq ft);
L_D = length of the wetland channel, m (ft); and
Prt = describe any pretreatment provided ahead of the channel (such as detention).

There are no emptying time parameters to report for wetland channels.

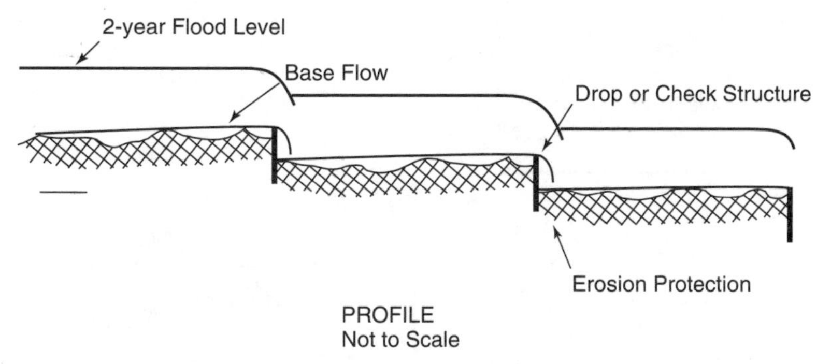

Figure 3.1 **Profile of an idealized wetland bottom channel (UDFCD, 1992).**

PARAMETERS FOR SAND FILTERS. Sand filters can be installed as basins or as sand filter inlets. Figures 3.2 through 3.4 illustrate typical sand filter designs. These installations will have a detention basin or a retention pond (or tank) upstream of the filter to remove the heavier sediment and, if properly designed, some of the oil and grease found in stormwater. However, such a pretreatment basin is not always present. All of the parameters required for a retention pond or for an extended detention basin should also be reported, along with the information about the sand filter whenever the filter is preceded by a pretreatment basin. For example, a filter inlet is often equipped with an underground tank, which helps to remove some of the sediment, oil, and grease before stormwater is applied to the filter. Such a tank is similar to a retention pond, and all of the parameters associated with a retention pond, such as volume, surface area, length, and surcharge volume, should be reported.

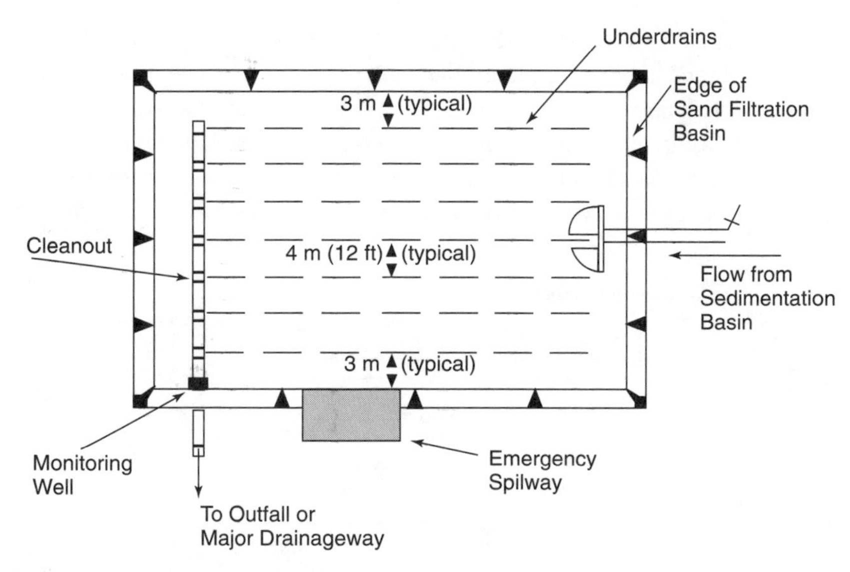

Figure 3.2 **Plan of an idealized sand filter basin.**

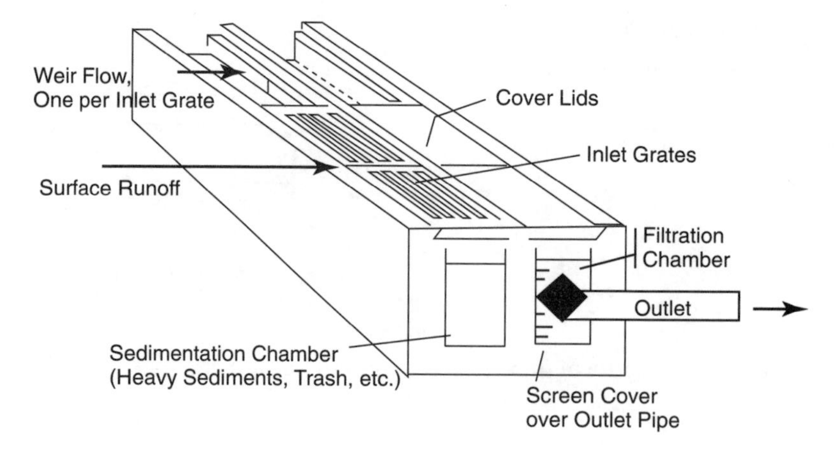

Figure 3.3 An idealized sand filter inlet.

In addition to the parameters of the pond or basin associated with the filter, provide the following:

- Dimensions of the installation;
- Depth of various filter material layers;
- Type of filter media, its median particle size (D_{50}), and its coefficient of uniformity;
- Maintenance frequency; and
- All associated drainage and flooding problems attributed to the installation because of its configuration size and maintenance practices.

Use the same parameters for *emptying time* as defined for the retention pond.

PARAMETERS FOR OIL, GREASE, AND SAND TRAPS. An oil, grease, and sand trap is an underground tank, similar to the one illustrated in Figure 3.4. It is nothing more then a special configuration of a retention pond. As a result, all parameters listed for a *retention pond* should be reported for these installations. Typically, these installations have a forebay and an outlet basin. In addition to reporting the parameters for a pond, pro-

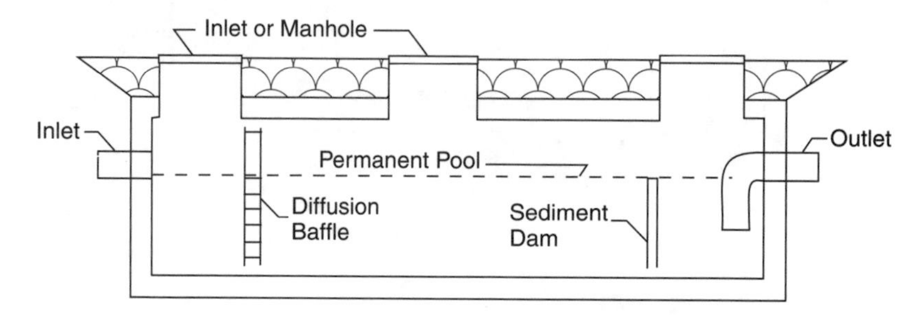

Figure 3.4 An idealized oil, grease, and sand trap.

vide the dimensions of the installation, details of its design (including skimmers, sorbent pillows, lamella plates, and baffles), and the maintenance provided during the testing period. Because these traps are much smaller than a surface pond, the flow-through velocity is of concern because it can cause trapped oil, grease, and sediment to be remobilized and flushed out of the trap. As a result, provide the average flow velocity that can be expected to occur in this device during a 2-year storm. This velocity can be used as an index for comparing the performance among a variety of installations. Use the same *emptying time* parameters as defined for a retention pond.

PARAMETERS FOR INFILTRATION AND PERCOLATION FACILITIES. For percolation trenches and for infiltration basins, report all of the parameters suggested for the extended detention basin. In addition, report the following:

- Depth to high groundwater and impermeable layers below the infiltrating surface of the basin or below the bottom of the percolation trench;
- The hydraulic conductivity of soils adjacent to percolation trenches and the saturated surface infiltration rates of soils underlying infiltration basins;
- Dimensions of the installation;
- Maintenance needs and associated drainage and flooding problems attributed to the installation; and
- Failures to empty out the captured water completely within the design emptying time.

Use the same *emptying time* parameters as defined for a retention pond.

SUMMARY

In summary, in the reporting of various BMP parameters and the field testing data on their performance, we can find a model for the selection of parameters useful in both more general monitoring programs and modeling. Table 3.3 lists the parameters identified in standardized procedures for BMP performance monitoring.

Table 3.3 Summary of reportable best management practice site parameters.

Parameter	Ret. pond	Ext. det. basin	Wet-land basin	Wet-land channel	Sand filter	Oil/ sand trap	Infilt. & perc.
Tributary watershed area—A_T	Yes	Yes	Yes	Yes	Yes	Yes	Yes
Total % trib. watershed is impervious—I_{IT}	Yes	Yes	Yes	Yes	Yes	Yes	Yes
% of impervious area hyd. connected—I_{IC}	Yes	Yes	Yes	Yes	Yes	Yes	Yes
Gutter/sewer/swale/ditches in watershed?	Yes	Yes	Yes	Yes	Yes	Yes	Yes
Average storm runoff volume—V_R	Yes	Yes	Yes	Yes	Yes	Yes	Yes
50th percentile runoff volume—V_{R50}	Yes	Yes	Yes	Yes	Yes	Yes	Yes
Coeff. var. of runoff volumes—CV_{VR}	Yes	Yes	Yes	Yes	Yes	Yes	Yes
Av. daily base flow volume—V_B	Yes	Yes	Yes	Yes	Yes	Yes	Yes
Average runoff interevent time—T_S	Yes	Yes	Yes	Yes	Yes	Yes	Yes
50th percentile interevent time—T_{S50}	Yes	Yes	Yes	Yes	Yes	Yes	Yes
Coeff. var. of interevent times—CV_{TS}	Yes	Yes	Yes	Yes	Yes	Yes	Yes
Average storm duration—T_D	Yes	Yes	Yes	Yes	Yes	Yes	Yes
50th percentile storm duration—T_{D50}	Yes	Yes	Yes	Yes	Yes	Yes	Yes
Coeff. var. of storm durations—CV_{TD}	Yes	Yes	Yes	Yes	Yes	Yes	Yes
Water temperature	Yes	Yes	Yes	Yes	Yes	Yes	Yes
Alkalinity, hardness, & pH	Yes	Yes	Yes	Yes	Yes	Yes	Yes
Sediment settling velocity dist.—V_{SD}	Yes	Yes	Yes	Yes	Yes	Yes	Yes
Type and frequency of maintenance	Yes	Yes	Yes	Yes	Yes	Yes	Yes
Inlet and outlet dimensions and details	Yes	Yes	Yes	Yes	Yes	Yes	Yes
Solar radiation	Yes	No	Yes	Yes	No	No	No
Volume of permanent pool—V_P	Yes	No	Yes	No	Yes	Yes	No
Perm. pool surface area—A_P	Yes	No	Yes	No	Yes	Yes	No
Littoral zone surface area—A_L	Yes	No	No	No	No	No	No
Length of permanent pool—L_P	Yes	No	Yes	No	Yes	Yes	No
Detention (or surcharge) vol.—V_D	Yes	Yes	Yes	No	Yes	Yes	Yes
Detention basin's surface area—A_D	Yes	Yes	Yes	No	Yes	Yes	Yes
Length of detention basin—L_D	Yes	Yes	Yes	No	Yes	Yes	Yes
Brim-full emptying time—T_E	Yes	Yes	Yes	No	Yes	Yes	Yes
1/2 Brim-full emptying time—$T_{0.5E}$	Yes	Yes	Yes	No	Yes	Yes	Yes
Bottom stage volume—V_B	No	Yes	No	No	No	No	No
Bottom stage surface area—A_B	No	Yes	No	No	No	No	No
Forebay volume—V_F	Yes	Yes	Yes	No	Yes	Yes	Yes
Forebay length—L_F	Yes	Yes	Yes	No	Yes	Yes	Yes
Wetland type, rock filter present?	No	No	Yes	Yes	No	No	No
% of wetland surface at $P_{0.3}$ & $P_{0.6}$ depths	No	No	Yes	Yes	No	No	No
Meadow wetland surface area—A_M	No	No	Yes	Yes	No	No	No
Plant species and age of facility	Yes	Yes	Yes	Yes	No	No	No
Two-year flood peak velocity	No	No	No	Yes	No	Yes	No
Depth to groundwater or impermeable layer	No	Yes	Yes	No	No	No	Yes

References

Ambrose, R.B., *et al.* (1988) Waste Allocation Simulation Models. *J. Water Pollut. Control Fed.*, **60**, 9, 1646.

Ambrose, R.B., and Barnwell, T.O., Jr. (1989) Environmental Software at the U.S. Environmental Protection Agency's Center for Exposure Assessment Modeling CEAM. U.S. EPA, Athens, Ga.

American Public Health Association (1995) *Standard Methods for the Examination of Water and Wastewater.* 19th Ed., Washington, D.C.

American Public Works Association (1969) Water Pollution Aspects of Urban Runoff. Rep. 11030DNS01/69, Fed. Water Pollut. Control Admin., Washington, D.C.

Barnwell, T.O., Jr. (1984) EPA's Center for Water Quality Modeling. *Proc. Int. Conf. Urban Storm Drainage.* Vol. 2., Chalmers Univ., Goteborg, Swed.

Barnwell, T.O., Jr. (1987) EPA Computer Models are Available to All. *Water Qual. Int.,* **2,** 19.

Bedient, P.B., and Huber, W.C. (1989) *Hydrology and Floodplain Analysis.* Addison-Wesley Publishers, Reading, Mass.

Bensen, M.A. (1962) Spurious Correlation in Hydraulics and Hydrology. *J. Hydraul. Eng.,* **91**, 57.

Bowie, G.L., *et al.* (1985) *Rates, Constants, and Kinetics Formulations in Surface Water Quality Modeling*. EPA-600/3-85-040, U.S. EPA, Athens, Ga.

Brown, R.G. (1984) Relationship between Quantity and Quality of Storm Runoff and Various Watershed Characteristics in Minnesota, USA. *Proc. 3rd Int. Conf. Urban Storm Drainage,* **3,** Chalmers Univ., Goteborg, Swed.

Di Toro, D.M. (1984) Probability Model of Stream Quality Due to Runoff. *J. Environ. Eng.,* **110,** 3, 607.

Di Toro, D.M., and Small, M.J. (1979) Stormwater Interception and Storage. *J. Environ. Eng.,* **105,** 43.

Driscoll, E.D. (1979) In *Benefit Analysis for Combined Sewer Overflow Control.* Seminar Publication, EPA-625/4-79-013, U.S. EPA, Cincinnati, Ohio.

Driscoll, E.D. (1981) *Combined Sewer Overflow Analysis Handbook for Use in 201 Facility Planning*. Final Rep., U.S. EPA, Facility Requirements Div., Policy and Guidance Branch, Washington, D.C.

Driscoll, E.D. (1986) Lognormality of Point and Nonpoint Source Pollutant Concentrations. In *Proceedings Stormwater Water Quality Users Group Meeting, Orlando, Florida*. EPA-600/0-86-023, U.S. EPA, Athens, Ga.

Driscoll, E.D., *et al.* (1989) *Pollutant Loadings and Impacts from Highway Stormwater Runoff.* Vol. I, Design Procedure (FHUA-RD-88006), and Vol. III, Analytical Investigation and Research Report (FHWA-RD-

99008), Office Eng. Highway Oper. Res. Develop., Fed. Highway Admin., McLean, Va.

Driver, N.E. and Tasker, G.D. (1988) Techniques for Estimation of Storm-Runoff Loads, Volumes, and Selected Constituent Concentrations in Urban Watersheds in the United States. U.S. Geol. Surv. Open-File Rep. 88-191, Denver, Colo.

Driver, N.E., *et al.* (1985) U.S. Geological Survey Urban Stormwater Data Base for 22 Metropolitan Areas Throughout the United States. U.S. Geol. Surv. Open File Rep. 85-337, Lakewood, Colo.

Feldman, A.D. (1981) *HEC Models for Water Resources System Simulation: Theory and Experience Advances in Hydroscience.* Vol. 12, Academic Press, New York, N.Y.

Fischer, H.B., *et al.* (1979) *Mixing in Inland and Coastal Waters.* Academic Press, New York, N.Y.

French, R.H., and McCutcheon, S.C. (Eds.) (1989) *Water Quality Modeling.* CRC Press, Boca Raton, Fla.

Gilbert, R. O. (1987) *Statistical Methods for Environmental Pollution Monitoring.* Van Nostrand Reinhold Co., New York, N.Y.

Gurtz, M.E. (1986) Development of a Research Data Management System: Factors to Consider. In *Research Data Management in the Ecological Sciences.* W.K. Michener (Ed.), Univ. S.C. Press.

Heaney, J.P., *et al.* (1975) *Urban Stormwater Management Modeling and Decision-Making.* EPA-670/2-75-022, U.S. EPA, Cincinnati, Ohio.

Herricks, E.E., and Schaeffer, D.J. (1987) Selection of Test Systems for Ecological Analysis. *Water Sci. Technol.,* **19,** 11, 47.

Holley, E.R, and Jirka, G.H. (1986) Mixing in Rivers. Tech. Rep. E-86-11, Corps Eng. Waterways Exp. Stn., Vicksburg, Miss.

Howard, C.D.D. (1976) Theory of Storage and Treatment Plant Overflows. *J. Environ. Eng.,* **102**, 709.

Huber, W.C. (1980) Urban Wasteload Generation by Multiple Regression Analysis of Nationwide Urban Runoff Data. In *Proc. Workshop Verif. Water Qual. Models.* R.V. Thomann and T.O. Barnwell (Eds.), EPA-600/9-80-016, U.S. EPA, Athens, Ga.

Huber, W.C. (1985) *Deterministic Modeling of Urban Runoff Quality in Urban Runoff Pollution.* H.C. Torno *et al.* (Eds.), NATO ASI Series, Series G: Ecological Sciences, Vol. 10, Springer-Verlag, New York, N.Y.

Huber, W.C. (1986) Modeling Urban Runoff Quality: State of the Art. *Proc. Conf. Urban Runoff Quality, Impact and Quality Enhancement Technol.,* B. Urbonas and L.A. Roesner (Eds.), Eng. Found. Am. Soc. Civ. Eng., New York, N.Y.

Huber, W.C., and Dickinson, R.E. (1988) *Storm Water Management Model User's Manual.* Version 4, EPA-600/3-88-00la, U.S. EPA, Athens, Ga.

Huber, W.C., and Heaney, J.P. (1982) *Analyzing Residuals Generation and Discharge from Urban and Nonurban Land Surfaces in Analyzing Natural Systems, Analysis for Regional Residuals—Environmental Quality*

Management. D.J. Basta and B.T. Bower (Eds.), Resour. for the Future, Johns Hopkins University Press, Baltimore, Md.

Huber, W.C., *et al.* (1982) *Urban Rainfall-Runoff-Quality Data Base*. EPA-600/2-81-238, U.S. EPA, Cincinnati, Ohio.

Hydroscience, Inc. (1979) *A Statistical Method for Assessment of Urban Stormwater Loads—Impacts—Controls*. EPA-440/3-79-023, U.S. EPA, Washington, D.C.

James, L.D., and Burges, S.J. (1982) *Selection, Calibration, and Testing of Hydrologic Models in Hydrologic Modeling of Small Watersheds*. C.T. Haan *et al.* (Eds.), Monograph No. 5, Am. Soc. Agric. Eng., St. Joseph, Mich.

Kibler, D.F. (Ed.) (1982) *Urban Stormwater Hydrology*. American Geophysical Union, Water Resour. Monograph 7, Washington, D.C.

Krenkel, P.A., and Novotny, V. (1980) *Water Quality Management*. Academic Press, New York, N.Y.

Lager, J.A., *et al.* (1977) *Urban Stormwater Management and Technology: Update and Users' Guide*. EPA-600/8-77-014, U.S. EPA, Cincinnati, Ohio.

Lijklema, L., *et al.* (Eds.) (1993) Interactions Between Sewers, Treatment Plants and Receiving Waters in Urban Areas: A Summary of INTERURBA '92 Workshop Conclusions. *Water Sci. Technol.,* **27**, 12.

Loganathan, V.G., and Delleur, J.U. (1984) Effects of Urbanization on Frequencies of Overflows and Pollutant Loadings from Storm Sewer Overflows: A Derived Distribution Approach. *Water Resour. Res.,* **20,** 7**,** 857.

McElroy, A.D., *et al.* (1976) *Loading Functions for Assessment of Water Pollution from Non-Point Sources*. EPA-600/2-76-151, U.S. EPA, Washington, D.C.

Manning, M.J., *et al.* (1977) *Nationwide Evaluation of Combined Sewer Overflows and Urban Stormwater Discharges—Vol. III: Characteristics of Discharges*. EPA-600/2-77-064c, U.S. EPA, Cincinnati, Ohio.

Metcalf and Eddy, Inc., *et al.* (1971) *Storm Water Management Model, Volume I—Final Report*. Rep. 11024DOC07/71, U.S. EPA, Washington, D.C.

Miller, R.A., *et al.* (1978) Statistical Modeling of Urban Storm Water Processes, Broward County, Florida. *Proc. Int. Symp. Urban Storm Water Manage.,* Univ. Kentucky, Lexington.

Mills, W.B., *et al.* (1985) *Water Quality Assessment: A Screening Procedure for Toxic and Conventional Pollutants in Surface and Ground Water* (Revised 1985). Parts 1 and 2, EPA-600/6-85-002a,b, U.S. EPA, Athens, Ga.

Noel, D.D., and Terstriep, M.L. (1982) Q-ILLUDAS—A Continuous Urban Runoff/Washoff Model. *Proc. Int. Symp. Urban Hydrol., Hydraul., and Sediment Control*, UKY BU128, Univ. Kentucky, Lexington.

Noel, D.D., *et al.* (1987) Nationwide Urban Runoff Program Data Reports. Illi. State Water Surv. Dep. Energy Natural Resour., Champaign, Ill.

Novotny, V. (1985) Discussion of Probability Model of Stream Quality Due to Run off by D.M. Di Toro. *J. Environ. Eng.,* **111,** 736.

Omemik, J. (1987) Ecoregions of the Conterminous United States. *Ann. Assoc. Am. Geogr.,* **77**, 118.

Pisano, W.C., and Queiroz, C.S. (1977) *Procedures for Estimating Dry Weather Pollutant Deposition in Sewerage Systems.* EPA-600/2-77-120, U.S. EPA, Cincinnati, Ohio.

Reckhow, K.H., and Chapra, S.C. (1983) *Engineering Approaches for Lake Management.* Butterworth Publishers, Woburn, Mass.

Rhoads, B.L. (1995) Stream Power: A Unifying Theme for Urban Fluvial Geomorphology. In *Stormwater Runoff and Receiving Systems.* E.E. Herricks (Ed.), Lewis Publishers, Boca Raton, Fla.

Roesner, L.A., and Dendrou, S.A. (1985) Discussion of Probability Model of Stream Quality Due to Runoff by D.M. Di Toro. *J. Environ. Eng.,* **111,** 5, 738.

Sartor, J.D., and Boyd, G.B. (1972) *Water Pollution Aspects of Street Surface Contaminants.* EPA-R2/72-081, U.S. EPA, Washington, D.C.

Schaeffer, D.J., *et al.* (1985) The Environmental Audit: I. Concepts. *Environ. Manage.,* **9***,* 191.

Small, M.J., and Di Toro, D.M. (1979) Stormwater Treatment Systems. *J. Environ. Eng.,* **105**, 557.

Tasker, G.D., and Driver, N.E. (1988) Nationwide Regression Models for Predicting Urban Runoff Water Quality at Unmonitored Sites. *Water Res. Bull.,*1091.

Thomann, R.V., and Mueller, J.A. (1987) *Principles of Surface Water Quality Modeling and Control.* Harper and Row, New York, N.Y.

Urbonas, B.R. (1995a) Recommended Parameters to Report with BMP Monitoring Data. *J. Water Res. Plann. Manage.,* **121,** 1, 23.

Urbonas, B.R. (1995b) Parameters to Report with BMP Monitoring Data. In *Stormwater NPDES Related Monitoring Needs.* H.C. Torno (Ed.), Am. Soc. Civ. Eng., New York, N.Y.

Urban Drainage and Flood Control District (1992) *Urban Storm Drainage Criteria Manual. Volume 3—Best Management Practices, Stormwater Quality.* Denver, Colo.

U.S. Environmental Protection Agency (1973a) *Biological Field and Laboratory Methods for Measuring the Quality of Surface Waters and Effluents.* EPA-670/4-73-001.

U.S. Environmental Protection Agency (1973b) *Methods for Identifying and Evaluating the Nature and Extent of Non-Point Sources of Pollutants.* EPA-430/9-73-014, Washington, D.C.

U.S. Environmental Protection Agency (1976a) *Land Use—Water Quality Relationship.* WPD-3-76-02, Water Plann. Div., Washington, D.C.

U.S. Environmental Protection Agency (1976b) *Areawide Assessment Procedures Manual.* EPA-600/9-76-014, Cincinnati, Ohio.

U.S. Environmental Protection Agency (1983) *Results of the Nationwide Urban Runoff Program.* Vol. I, Final Rep., Washington, D.C.

U.S. Environmental Protection Agency (1989) *Rapid Bioassessment Protocols for Use in Streams and Rivers—Benthic Macroinvertebrates and Fish.* EPA 440-489-001.

U.S. Environmental Protection Agency (1991) *Technical Support Document for Water Quality-Based Toxics Control.* EPA-505/2-90-001.

Viessman, W., *et al.* (1989) *Introduction to Hydrology.* 3rd. Ed., Harper and Row, New York, N.Y.

Walker, J.F., *et al.* (1989) Spreadsheet Watershed Modeling for Nonpoint-Source Pollution Management in a Wisconsin Area. *Water Res. Bull.*, **25**, 1, 139.

Water Pollution Control Federation (1989) *Combined Sewer Overflow Pollution Abatement.* Manual of Practice No. FD-17, Alexandria, Va.

Whipple, D.J., *et al.* (1983) *Stormwater Management in Urbanizing Areas.* Prentice-Hall, Englewood Cliffs, N.J.

Wischmeier, W.H., and Smith, D.D. (1958) Rainfall Energy and Its Relationship to Soil Loss. *Trans. Am. Geophys. Union,* **39**, 2, 285.

Woodward-Clyde Consultants (1989) Synoptic Analysis of Selected Rainfall Gages Throughout the United States. Rep. to U.S. EPA, Oakland, Calif.

Zison, S.W. (1980) Sediment-Pollutant Relationships in Runoff from Selected Agricultural, Suburban and Urban Watersheds. EPA-600/3-80-022, U.S. EPA, Athens, Ga.

Zukovs, G., *et al.* (1986) Development of the HAZPRED Model. In *Proceedings Stormwater Water Quality Model Users Group Meeting, Orlando, Florida.* EPA-600/9-86-023, U.S. EPA, Athens, Ga.

SUGGESTED READINGS

Environment Canada (1983) *Sampling for Water Quality.* Water Qual. Branch, Inland Waters Directorate, Ottawa, Can.

U.S. Environmental Protection Agency (1985) *Water Quality Assessment: A Screening Procedure for Toxic and Conventional Pollutants.* Parts I and II, EPA-600/6-85-002a and 002b.

U.S. Environmental Protection Agency (1986) *Stream Sampling for Waste Load Allocation Applications Handbook.* EPA-625/6-86-013.

U.S. Environmental Protection Agency (1992) *Biological Criteria: Technical Guidance for Streams and Small Rivers.* Draft.

U.S. Environmental Protection Agency (1993) *Biological Field and Laboratory Methods for Measuring the Quality of Surface Waters and Effluents.* EPA-670/4-73-001.

U.S. Geological Survey (1993) *Guidelines for the Processing and Quality Assurance of Benthic Invertebrate Samples Collected as Part of the National Water-Quality Assessment Program.* Open-File Rep. 93-407.

Chapter 4
Source Controls

Source controls are practices that prevent pollution by reducing potential pollutants at their source before they come into contact with stormwater, as opposed to treatment controls that remove pollutants from stormwater. A comprehensive urban runoff quality management program requires that certain source control best management practices (BMPs) be implemented on existing development. In addition, some source controls will be applied in developing areas after the new development has been completed. Twenty-two source control BMPs are described in this chapter.

Typically, source controls for urban areas can be grouped into the following seven categories:

- Public education—this is an institutional practice intended to change the way the general public manages many of the constituents that wind up in stormwater runoff. How the public uses and disposes of automotive fluids, fertilizers, pesticides, herbicides, and many other household products can have a profound effect on the quantities of these materials that come into contact with stormwater and the amounts of these substances that are eventually discharged to the receiving waters. Although this promises to be a cost-effective way of affecting stormwater quality, the effectiveness of public education on the actual reductions of the target constituents in receiving waters has yet to be definitively demonstrated.

- Planning and management of developing areas—these practices by local governments can be aimed at reducing runoff and the discharge of pollutants through stormwater from new developments and are most effective when applied during the site-planning phase of new development. Examples include the adoption of zoning ordinances and subdivision regulations aimed at stormwater quality management. These ordinances may require buffers and setbacks from all streams, lakes, and natural wetlands and may include provisions to reduce impervious areas that are connected directly to the formal stormwater drainage system.
- Materials management—these practices include controlling the use, storage, and disposal of chemicals that could pollute runoff. The objective is to reduce the opportunity for rainfall or runoff to come into contact with these chemicals. This BMP includes the following three categories of activities:
  - Material use controls,
  - Material exposure controls, and
  - Material disposal and recycling controls.
- Spill prevention and cleanup—this category includes programs that reduce the risk of spills during outdoor handling and the transportation of chemicals and other materials and the development of plans and programs to respond, contain, and rapidly clean up spills when they do occur so that they do not enter the storm drain system.
- Illegal dumping controls—this category comprises ordinances, public education programs, and enforcement aimed at keeping individuals and businesses from dumping various waste products onto the urban landscape and the drainage system.
- Street/storm drain maintenance—this applies to the removal of pollutants from paved areas and the maintenance of runoff quality controls that exist within the drainage system. Examples include street sweeping, catch basin cleaning, road and bridge maintenance, and maintenance of structural controls in the system for runoff quality management. This group also includes the use of good housekeeping measures whenever performing pavement maintenance such as asphalt overlays or seal and chip procedures.
- Illicit connection controls—this group of controls is directed at preventing, by ordinance, and eliminating, by discovery and removal, connections to the storm drainage system that discharge any material except stormwater runoff. Bans on connection of floor drains, washdown areas, septic tank overflows, and the like to the stormwater conveyance system are all a part of this BMP category.

GENERAL GUIDANCE FOR SELECTION OF STORMWATER BEST MANAGEMENT PRACTICES

Selecting the proper stormwater quality controls or BMPs is often driven by the following:

- Federal, state, and local regulations;
- Real or perceived receiving water problems or beneficial uses to be protected; or
- The cost of the BMPs being considered.

The reduction of pollutants in stormwater discharges to the maximum extent practicable (MEP) is the statutory requirement of stormwater regulations. Ultimately, however, the goal is to reduce the effects of urban stormwater runoff on the receiving water. Despite the best of intentions, the cost of the selected BMPs is a major consideration, especially when considering retrofitting treatment control BMPs in developed areas. Retrofitting any treatment control BMP is expensive and often unfeasible on a citywide basis. For this reason, source controls are often the only affordable option.

There is no single BMP that will prevent all of the effects on receiving waters caused by urban runoff. However, through the use of a combination of BMPs, both source controls and treatment controls, the greatest benefits will be gained. Figure 2.2 illustrates a multilevel strategy suggesting that urban runoff quality management begins with source controls, followed by on-site treatment controls. These can be further supplemented, where required, by regional, subregional, or communitywide control facilities. This so-called "treatment train" (Livingston *et al.*, 1988) can use the numbers and types of controls that best serve the community or the watershed. Land availability, capital, and operation and maintenance (O & M) costs, balanced against the benefits of pollutant removal, should be the bases for determining the nature of such a treatment train at each site.

SELECTING SOURCE CONTROLS. Table 4.1, with locally defined selection criteria for various factors, contains a worksheet that can be used to assist a municipality in selecting source controls. Several municipalities have used this system or a similar one to select BMPs (City of Stockton, 1993, and County of San Bernardino, 1993). Some of the locally defined selection criteria may include

Table 4.1 Worksheet for evaluating municipal source control practices (Camp Dresser & McKee *et al.,* 1993).

Worksheet 1
Source control practices

Program activities: (such as residential/commercial), See Table 4.2
Program element: (such as roadway and drainage facility maintenance), see Table 4.2

Practice	Meets regulatory requirements (1–5)	Effectiveness of pollutant removal (1–5)	Public acceptance (1–5)	Implementable (1–5)	Institutional contraints (1–5)	Costs (1–5)	Total (30 maximum)

- Ability to meet regulatory requirements,
- Effectiveness of the practice to remove pollutants of concern,
- Public acceptance of the practice,
- Ability to implement,
- Institutional constraints, and
- Costs.

Selection criteria provide for a sliding scale of 1 to 5 and can be used to rank the practices in how well they meet the factors or concerns represented by the criteria. To use the Table 4.1 worksheet, first determine which proposed stormwater program elements are addressed by the various management practices or source control BMPs previously listed. Table 4.2 provides an example of how this can be done. For each BMP identified, the municipality ranks it according to its ability to meet selection criteria.

After the worksheets are completed for each program element and the ratings are scored, the municipality will have a ranking of BMPs. It should be kept in mind that the ranking is only a tool for comparing BMPs and provides the information for the municipality to decide which BMPs should be implemented immediately, which BMPs should be targeted for pilot-scale study, and which BMPs should be phased for later implementation. Such ranking may also serve the purpose of defining MEP as required under stormwater regulations.

Selection criteria and the scoring system presented in this chapter are similar to other qualitative selection processes developed to screen and rank alternatives for stormwater management programs. The user may wish to modify the selection process to accommodate local requirements. Modification of the following selection process attributes may be considered:

- Criteria—the user may want to redefine some of the criteria or add or subtract criteria.
- Scores—likewise, the user may want to modify the scoring to a simple +, 0, and/or 1, 2, and 3.
- Weighing—in addition, it may be appropriate to group the criteria into tiers reflecting their relative importance to solid waste management practice goals. By multiplying the scores of the highest tier by some factor (for example, ×2) the first-tier scores could be weighted more heavily than the others to reflect this importance.
- Fatal flaw—scoring the BMPs should provide for some fatal flaw (for example, the BMP is illegal or its implementation is unacceptable to the public) that would make implementation impossible. Scoring a fatal flaw as a "0" is one way of highlighting the flaw. Any BMP scoring a "0" against a criterion would be eliminated from consideration, regardless of its overall ranking.

Ranking Criteria. A suggested criteria for ranking source control BMPs follows.

MEETS REGULATORY REQUIREMENTS. Does the BMP comply with federal stormwater regulations or a state permit condition? For the most part, the selected BMP will address the requirements of the stormwater regulations. In certain situations, the state agency may require a specific BMP, in which case it will become a mandated best management practice.

Rating score:

5 = Meets specific state requirements.
3 = Meets federal storm water regulations.
1 = Does not meet regulatory requirements.

EFFECTIVENESS OF POLLUTANT REMOVAL. Does the BMP have a high likelihood of reducing pollutants of concern? This is probably one of the hardest questions to answer, especially for source control BMPs. As of 1996, the knowledge required to make this assessment is lacking. Consequently, most source control BMPs will receive a low rating, not so much because they are not effective in removing pollutants, but because the ability to quantify the removal is not available. It is more likely that the ratings of source control BMPs will be relative to each other. Also, some BMPs are more suited to removing a specific pollutant than others.

Table 4.2 Application of source control practices to stormwater management plan program elements (Camp Dresser & McKee *et al.*, 1993).[a]

Required elements of solid waste management program	Source control practice							
	Planning management	Material use control	Material exposure controls	Material disposal and recycling	Spill prevention and cleanup	Illegal dumping controls	Illicit connection controls	Street/storm drain maintenance
For residential/commercial activities:								
Roadway and drainage facility maintenance		X						X
Best management practice planning for new development and redevelopment projects	X							X
Retrofitting existing or proposed flood control projects with best management practices								

Municipal waste handling and disposal operations	X	X	X	X	X	
Pesticide, herbicide, and fertilizer use controls	X	X	X	X	X	
For improper discharge activities:						
Prevention, detection, and removal of illegal connections to storm drains						X
Spill prevention, containment, and response	X	X	X	X	X	
Promote proper use and disposal of toxic materials	X	X	X	X	X	
Reduce stormwater contamination by leaking/overflowing separate sanitary sewers				X	X	

Rating score:

5 = Highly effective in removing pollutants with sufficient data to support such a claim.

3 = Expected to provide moderate level of pollutant removal.

1 = Ineffective in removing pollutants or insufficient data are available to make an assessment.

PUBLIC ACCEPTANCE. Does the BMP have public support? Some source control BMPs will carry more public support than others (for example, stream cleanup versus tighter land-use controls). The successful implementation of source controls depends, to a large extent, on the amount of public support. Such support can be identified by knowing the community interests. Without the public support (which should include understanding the issues and problems), the BMP will be ineffective.

Rating score:

5 = Public understands the problem and supports the BMP implementation.

3 = Likely that the public will support the BMP once it understands the problem but presently does not know the issues.

1 = Public does not support the BMP.

IMPLEMENTABLE. Can the BMP be implemented through existing programs or departments? The ability of the municipality to implement a BMP will, to a certain extent, depend on whether existing programs can be used or expanded. Obviously, the likelihood of a BMP being implemented is greatest if it can be incorporated to an existing program (or department). Another issue to be considered under this criterion is the availability of staff (or necessity of additional staff) and equipment. Also, the municipality should consider how the BMP will affect interdepartmental coordination and communication. (Is there overlap? Will there be "turf battles"?)

Another issue to consider under implementation is whether the BMP should be implemented in a specific area or a larger watershed area. Some BMPs will be more appropriate for a specific target group within a limited area (for example, the industrial illicit connection program), while others need to be implemented areawide (for example, the elimination of motor oil dumping). Those BMPs that apply to the larger watershed should receive higher consideration.

Rating score:

5 = Existing program or department can be used and adequate personnel and equipment are available; applies to larger watershed area.

3 = Existing program will need to be expanded with either more staff or equipment.

1 = Existing program or department does not exist to implement BMPs.

INSTITUTIONAL CONSTRAINTS. Are there any institutional constraints that limit the ability to implement the BMP? Typical institutional constraints would include legal and intergovernmental coordination. Many of the source control BMPs can be implemented under existing ordinances or regulations. A new ordinance will be required in some situations. Source control BMPs may also require interjurisdictional coordination, to be effective. This coordination, whether it is through complementary watershed protection ordinances, common public education programs, or shared maintenance duties, is critical to selecting BMPs.

Rating score:

5 = Existing ordinances and intergovernmental agreements are in place to implement BMP.

3 = New ordinances and intergovernmental agreements will need to be developed; however, there is consensus among the parties that the BMP is important.

1 = New ordinances and intergovernmental agreements will need to be developed. Such ordinances and agreements will require extensive time and cost to develop a consensus.

COST. How much is the BMP going to cost initially and over the long term, and does the municipality have adequate financial means to fund its implementation? Many source control BMPs do not require significant capital investments (such as storm drain stenciling), while others do (such as the purchase of vacuum street sweepers). Also, it is important to look at the means for generating the funds required for the BMP (or for the stormwater program in general). Is the existing funding mechanism (for example, user fees or general funds) adequate for funding the BMP, or does a new mechanism need to be developed (such as a utility fund)? Additionally, the municipality may want to consider the cost to the community at large, although this may be a difficult task.

Rating score:

5 = Low-cost BMPs that can be funded with the existing municipal funding mechanism.

3 = Moderate-cost BMPs that will require an adjustment to the city's funding mechanism for support.

1 = High-cost BMP or a BMP that will require a major restructuring of the municipal funding mechanism.

There are a few points to remember about selection processes:

- Have several people or a stormwater committee conduct the selection independently to get a broad perspective on the relative merits of each BMP and to help reach consensus.
- The validity and accuracy of any scoring system is only as good as the available information.
- Keep the selection system as simple as possible and use "best professional judgment" to interpret and conduct a reality check on the total scores. Differences of a few points in the total score are probably not significant.
- The final rankings may be used to plan and prioritize the stormwater management program. For example, those BMPs with the highest scores may be implemented in the first year of a new program, while low-scoring BMPs may need time for development, relegating their implementation to later years or to further study.
- The exercise of working through this selection will provide the necessary data to promote the stormwater program to other departments, political leaders, regulatory agencies, and the public.

PUBLIC EDUCATION AND PARTICIPATION

Public education and participation, like an ordinance or piece of equipment, is not so much a BMP as it is a method by which to implement BMPs. The Clean Water Act and the 1990 stormwater regulations require public participation and the establishment of public education programs. This section highlights the importance of integrating elements of public education and participation to a municipality's overall plan for urban runoff quality management. Public education and participation are vital components of many of the individual source control BMPs that follow in this chapter.

A public education and participation plan provides the municipality with a strategy for educating its employees, the public, and businesses about the importance of protecting stormwater from improper use, storage, and disposal of pollutants. Municipal employees must be trained, especially those who work in departments not directly related to stormwater but whose actions affect stormwater. Residents must become aware that a variety of hazardous products are used in the home and that their improper use and disposal can pollute stormwater. Businesses, particularly smaller ones that may not be regulated by federal, state, or local regulations, must be informed of ways to reduce their potential to pollute stormwater.

The specific public education and participation aspects of each of the source controls are highlighted in the sections for each BMP discussed in this chapter. The focus of this section includes the overall objectives and approaches for ensuring public involvement in local stormwater management programs.

OBJECTIVES. The public education and participation plan should be based on five objectives:

- Promote a clear identification and understanding of the problem and the solutions;
- Identify responsible parties and efforts to date;
- Promote community ownership of the problems and the solutions;
- Change behaviors; and
- Integrate public feedback to program implementation.

APPROACH. The approach to public education and participation is detailed in the following bulleted items:

- Pattern a new program after the many established programs from municipalities around the state and country. Whenever possible, integrate stormwater public education and participation to existing programs from other departments at the municipality.
- Implement public education and participation as a coordinated campaign in which each message is related to the last.
- Present a clear, consistent message and image to the public regarding how they contribute to stormwater pollution and what they can do to reduce it.
- Stick to the program. There is a lag in the public's response to anything new, so it is important to stick with the message long enough to get the bulk of the audience. The point in time when municipal staff are ready to "move on" to another message is probably about the time that most of the audience is just starting to get the original message.
- Expand the definition of "public" to include small businesses that often possess the same limited levels of awareness of the problems, regulations, and solutions as the general public. As a result, small busi-

nesses need the same level of technical assistance (education) and participation in the process as the general public.

- Use a multimedia approach to reach the full range of audiences.
- Translate messages to the foreign languages of the community to reach the full spectrum of the populace and avoid misinterpretation of messages. Account for cultural differences in translating messages and concepts. Outreach in a non-English language is not just a matter of transcription.
- Create an awareness and identification of the local watershed.
- Involve focus or advisory groups in the development of a public education and participation plan. This will create a more effective plan and promote ownership of the plan by those involved.
- Use everyday language in all public pieces. Use outside reviewers to highlight and reduce the use of technical terminology, acronyms, and jargon.
- Make sure all statements have a sound, up-to-date technical basis. Do not contribute to the spread of misinformation.
- Break up complicated subjects into smaller, more simple concepts. Present these concepts to the public in a simplified and organized way to avoid "overloading" and confusing the audience.
- Choose quality over quantity. One good message or outreach piece is more effective than many poor attempts.

ADMINISTRATIVE AND STAFFING CONSIDERATIONS. Because most of the BMPs discussed in this publication require some public education and participation, a qualified public education specialist can be critical to the success of source control programs.

EXAMPLES OF EFFECTIVE PROGRAMS. There are a number of communities with effective public education and participation programs. The most proactive include the Alameda Countywide Clean Water Program, the City and County of San Francisco, the Santa Clara Valley Nonpoint Source Pollution Control Program, and the City of Palo Alto, California; the Municipality of Metropolitan Seattle (Metro), Washington; and the Unified Sewerage Agency of Washington County, Oregon. In addition, large businesses, such as utility companies, have used inserts in their bill mailings to educate their customers.

LAND-USE PLANNING AND MANAGEMENT

This BMP presents an important opportunity to reduce the pollutants in stormwater runoff by using a comprehensive planning process to control or prevent certain land-use activities in areas where water quality is sensitive to development. It is applicable to all types of land use and represents one of the most effective pollution prevention practices. Land-use planning and management are critical to watershed management.

APPROACH. The land-use planning process need not be complex. A basic schematic model involves six basic phases as follows:

- Phase 1—goals: clear-cut water quality goals are determined.
- Phase 2—study: activities of this phase are identifying planning area, gathering pertinent data, and writing a description of the planning area and its associated problems.
- Phase 3—analysis and synthesis: the water quality goals are determined and prioritized as they relate to land use.
- Phase 4—recommendations: future courses of action are developed to address the identified problems and needs.
- Phase 5—adoption: recommendations are presented to a political body for acceptance and implementation.
- Phase 6—implementation: recommendations adopted by the local government are implemented by the community.

COST CONSIDERATIONS. The majority of the cost for this BMP is associated with establishing a comprehensive land-use plan that addresses the quality of stormwater runoff after projects are completed.

REGULATORY CONSIDERATIONS. Ordinances typically are required to implement and enforce land-use plans, including those relating to stormwater runoff quality. Several federal initiatives influence land-use planning, including the National Environmental Policy Act, the Clean Water Act, and the Clean Air Act.

ADMINISTRATIVE AND STAFFING CONSIDERATIONS. Site plans or environmental impact documents for projects must be reviewed for compliance. Additional staff may be required to implement a site plan review and inspection program. Also, interdepartmental and decisionmaker cooperation is crucial.

PUBLIC EDUCATION AND PARTICIPATION CONSIDERATIONS. To gain the necessary support for land-use policies, public participation is a necessity. The public should be educated regarding the positive environmental effects of land-use management, including the improvement to stormwater runoff quality. To increase public awareness, public education materials should be as specific as possible about the effects of land-use policies on watersheds and water quality. Geographic information systems can be a dynamic and effective tool to illustrate water quality effects.

LIMITATIONS. Land-use planning and management frequently address sensitive public issues. Restrictions on certain land uses required to mitigate stormwater pollution may not be politically feasible. Zoning ordinances that are not reinforced by a comprehensive planning process typically are less effective because they are often applied illogically for pollution prevention and are more easily circumvented politically.

EXAMPLES OF EFFECTIVE PROGRAMS. The city of Austin, Texas, has chosen to manage the effects of new development in two fundamental ways: through treatment control requirements and through impervious area minimization requirements. The city of Olympia, Washington, studied the feasibility of a 20% reduction in impervious surfaces throughout northern Thurston County, Washington, and found the goal achievable through policy changes, new standards, and education. In Virginia, the Chesapeake Bay Preservation Act requires no net increase in pollutants in stormwater runoff from previously undeveloped sites. Runoff from redeveloped sites must contain 10% fewer pollutants than before redevelopment.

HOUSEKEEPING PRACTICES

The promotion of efficient and safe housekeeping practices (storage, use, cleanup, and disposal) when handling potentially harmful materials such as fertilizers, pesticides, cleaning solutions, paint products, automotive products, and swimming pool chemicals can be an effective source control BMP. Good housekeeping practices include storing hazardous products securely, safely, and in original containers; reading and following product instructions; working in well-ventilated areas; and properly disposing of products.

Related information is provided in source control BMPs for safer alternative products, household hazardous waste collection, used oil recycling, and spill prevention and cleanup.

APPROACH. The following housekeeping practices may be effective:

- Pattern a new program after the many established programs from municipalities around the state and country. Integrate this BMP as much as possible with existing programs in the municipality.
- This BMP involves three key audiences: municipal employees, the general public, and small businesses.
- Implement this BMP in conjunction with the safer alternative products BMPs.

COST CONSIDERATIONS. The primary cost for good housekeeping practices is for staff time. More information follows under the heading Administrative and Staffing Considerations.

REGULATORY CONSIDERATIONS. There are no additional regulatory requirements to this BMP. Existing regulations already require municipalities to properly store, use, and dispose of hazardous materials and waste. This source control also focuses on materials and waste that may not be hazardous in the regulatory sense but are deleterious to water quality and organisms. Housekeeping practices of the general public are addressed through education rather than regulation.

ADMINISTRATIVE AND STAFFING CONSIDERATIONS. Staff are needed to train municipal employees and coordinate public education efforts. Municipal employees who handle potentially harmful materials should be trained in good housekeeping practices.

PUBLIC EDUCATION AND PARTICIPATION CONSIDERATIONS. Public awareness is a key to this BMP. The continued use or switch to good housekeeping practices is a behavior, and behavior is based on awareness.

LIMITATIONS. There are no major limitations to this BMP.

EXAMPLES OF EFFECTIVE PROGRAMS. There are a number of communities with effective programs. The most proactive include Santa Clara County, the city of Palo Alto, and the city and county of San Francisco, California, and the Municipality of Metropolitan Seattle (Metro), Washington. These programs are characterized by high-profile, comprehensive efforts to reach all audiences (that is, the general public, small businesses, and agency employees) using a variety of tools.

SAFER ALTERNATIVE PRODUCTS

Promoting the use of less harmful products can reduce the amount of toxic and deleterious substances that enter stormwater and ultimately reach receiving waters. Alternatives exist for most product classes, including fertilizers, pesticides, cleaning solutions, and automotive and paint products. There are natural alternatives to most garden products and less toxic alternatives to home and automotive repair products.

Related information is provided in source control BMPs for housekeeping practices, household hazardous waste collection, used oil recycling, and spill prevention and cleanup.

APPROACH. Pattern a new program after the many established programs from municipalities around the country. Integrate this BMP as much as possible with existing programs at the municipality. This BMP has three key audiences: municipal employees, the general public, and small businesses. Implement this BMP in conjunction with the housekeeping practices BMP.

COST CONSIDERATIONS. The primary cost of this BMP is for staff time. More information is available in the following sections.

REGULATORY CONSIDERATIONS. This BMP has no additional regulatory requirements. Existing regulations already require municipalities to reduce the use of hazardous materials. Safer alternatives for use by the general public are presented through education rather than required by regulation.

ADMINISTRATIVE AND STAFFING CONSIDERATIONS. Staff are needed to educate municipal employees and coordinate public education efforts. Municipal employees who handle potentially harmful materials should be trained in the use of safer alternatives. Purchasing departments should be encouraged to procure less hazardous materials.

PUBLIC EDUCATION AND PARTICIPATION CONSIDERATIONS. Awareness is the key to this BMP. It promotes a willingness to try alternatives and modify old behaviors.

LIMITATIONS. Safer alternative products may not be available, suitable, or effective in every case.

EXAMPLES OF EFFECTIVE PROGRAMS. There are a number of communities with effective programs promoting safer alternative products. The most proactive include Santa Clara County, the city of Palo Alto, and the city and county of San Francisco, California, and the Municipality of Metropolitan Seattle (Metro), Washington. The Bio-Integral Resource Center in Berkeley, California, conducts research and produces brochures and a newsletter on integrated pest management.

MATERIAL STORAGE CONTROL

Material storage controls can prevent or reduce the discharge of pollutants to stormwater from material delivery and storage areas. This can be done by reducing the storage of hazardous materials on site, storing materials in a designated area, installing secondary containment, conducting regular inspections, and training employees and subcontractors.

This BMP primarily concerns material delivery and storage for municipal and commercial operations. For material storage related to the general public (for example, the storage of pesticides), refer to the Housekeeping Practices section of this chapter.

APPROACH. The key is to design and maintain material storage areas that reduce exposure to stormwater by

- Storing materials inside or under cover on paved surfaces;
- Using secondary containment, where needed;
- Minimizing storage and handling of hazardous materials; and
- Inspecting storage areas regularly.

Keep an ample supply of spill cleanup materials near the storage area.

COST CONSIDERATIONS. Costs will vary depending on the size of the facility and the necessary controls.

REGULATORY CONSIDERATIONS. The storage of reactive, ignitable, or flammable liquids must comply with the Uniform Fire Code and the National Electric Code. The storage of reactive, ignitable, or flammable liquids and chemicals is regulated by Superfund Amendments and Reauthorization Act Title III, in excess of the minimum quantities set forth in the act.

ADMINISTRATIVE AND STAFFING CONSIDERATIONS. Accurate and up-to-date inventories should be kept of all stored materials. Employees should be well trained in proper material storage. Employee education is paramount for successful BMP implementation.

LIMITATIONS. Storage sheds often must meet building and fire code requirements.

VEHICLE-USE REDUCTION

Reducing the discharge of pollutants to stormwater from vehicle use can be achieved by highlighting the stormwater effects from vehicle emissions, promoting the benefits to stormwater of alternative transportation, and integrating initiatives with existing or emerging regulations and programs.

APPROACH. The following practices may be successful in implementing vehicle-use reduction BMPs:

- Build alliances with air quality agencies to identify common challenges and opportunities.
- Integrate this BMP as much as possible with efforts being developed and implemented by government agencies and businesses to reduce vehicle use and improve air quality, including the designation of high-occupancy vehicle or carpool lanes in most major cities in America. Integration will help avoid redundant or conflicting programs and will be more effective and efficient.

- Establish trip reduction programs at government offices or large businesses.

COST CONSIDERATIONS. The primary cost is for staff time.

REGULATORY CONSIDERATIONS. Support efforts to pass reasonable regulations at the state and local level (land-use plans and zoning ordinances) aimed at reducing vehicle use and developing transit-oriented communities. Also, support development of regional governing bodies to address the issue in a comprehensive way.

ADMINISTRATIVE AND STAFFING CONSIDERATIONS. This BMP requires at least one staff person to track, review, and comment on emerging legislation and programs.

PUBLIC EDUCATION AND PARTICIPATION CONSIDERATIONS. Educate the public and municipal employees about the water quality benefits of reduced vehicle use. Also, help coordinate public participation in ride-sharing programs.

LIMITATIONS. The limitations for this BMP include

- The possible lack of cooperation and integration between departments and programs, and
- The fact that the use of alternative transportation may be dependent on its convenience and relative cost.

STORM DRAIN SYSTEM SIGNS

Stenciling of the storm drain system (inlets, catch basins, channels, and creeks) with prohibitive language and graphic icons discourages the illegal dumping of unwanted materials.

APPROACH. Create a volunteer workforce to stencil storm drain inlets, and use municipal staff to erect signs near drainage channels and creeks. Enlist the aid of city code enforcement staff to stencil curb inlets that show signs of being used for dumping.

COST CONSIDERATIONS. The following bulleted items should be considered for this BMP:

- A volunteer workforce serves to lower program cost.
- Stenciling kits require procurement of durable and disposable items.
- The storage and maintenance of stenciling kits requires planning.

- The program should aid in the cataloging of the storm drain system.

REGULATORY CONSIDERATIONS. Develop, implement, and enforce an ordinance that requires inlets, catch basins, channels, and creeks to be fitted with antidumping, pollution prevention signs.

ADMINISTRATIVE AND STAFFING CONSIDERATIONS. The primary staff demand is for program setup to provide marketing and training. Ongoing/follow-up staff time is minimal because of volunteer services. A minimum of two persons is required for stenciling in high-traffic areas and commercial and industrial zones with appropriate safety measures in use (for example, reflective vests, flag person, and signage). Additional staff may be required at program headquarters for emergencies or to answer questions.

PUBLIC EDUCATION AND PARTICIPATION CONSIDERATIONS. Promote volunteer services (individual and business) through radio, television, and mail-out campaigns. Encourage public reporting of improper waste disposal by a hotline number stenciled onto the storm drain inlet. Training sessions of approximately 10 to 15 minutes will cover stenciling procedures, including how to stencil, recordkeeping, and problem drain notation. Also, consider proper health and safety protocols (such as the buddy system, traffic, and health concerns).

LIMITATIONS. The following limitations may apply.

- Private property access limits stenciling to publicly owned areas.
- This program is dependent on volunteer response.
- Storm drain inlets that are physically blocked will be missed or require follow-up.
- High-traffic, commercial, or industrial zones will be the responsibility of city staff.

EXAMPLES OF EFFECTIVE PROGRAMS. The city of Palo Alto, California, has a combined volunteer/contractor program that greatly facilitates storm drain stenciling. The city used the Conservation Corps to paint approximately 75% (2 000) of its storm drains, leaving the 25% (700) of its drains in more residential areas to be done by volunteers. This strategy speeds up the stenciling, reduces the city's liability, supports a worthwhile program, and still allows plenty of storm drains for volunteers.

The Association of Bay Area Governments in Oakland, California, has sponsored a nine-county stenciling effort on Earth Day since 1992. This association has up-to-date information on stencil and program development.

The city of Huntington Beach, California, has a stencil that includes the municipal code section number for illegal dumping to facilitate incident reporting and enforcement.

HOUSEHOLD HAZARDOUS WASTE COLLECTION

Household hazardous wastes (HHWs) are defined as waste materials that typically are found in homes or similar sources and exhibit characteristics such as corrosivity, ignitability, reactivity, and/or toxicity or are listed as hazardous materials by the U.S. Environmental Protection Agency (U.S. EPA). This source control also focuses on the collection of deleterious chemicals that sometimes are disposed of in a manner that threatens stormwater quality.

APPROACH. Integrate efforts with a municipal solid waste program that likely has already been established. Optimize collection method(s) (for example, permanent, periodic, mobile, and curbside) and frequency (for example, monthly and quarterly) based on waste type, community characteristics, existing programs, and budgets.

COST CONSIDERATIONS. The following cost considerations may apply to this BMP:

- Both collection and disposal can be expensive and are partly a function of the frequency of collection, which depends on the collection program implemented.
- Trained operators are required.
- Laboratory and detection equipment are necessary.
- Extensive recordkeeping is required including dates, types, and quantities.
- Many communities have deferred HHW programs because of the high cost.
- Cost depends on the type of program chosen and available disposal costs.

REGULATORY CONSIDERATIONS. Federal regulations (such as the Resource Conservation and Recovery Act; the Superfund Amendments and Reauthorization Act; and the Comprehensive Environmental Response, Compensation, and Liability Act) and state regulations regarding the disposal of hazardous waste apply to this BMP. Local ordinances to discourage improper disposal may be necessary. Municipalities may be required to have HHW elements within their integrated waste management plans.

ADMINISTRATIVE AND STAFFING CONSIDERATIONS. This BMP may require a minimum of six highly trained persons per collection site or event to handle traffic, waste drop-off, characterization, and disposal.

PUBLIC EDUCATION AND PARTICIPATION CONSIDERATIONS. The following considerations may be applicable for this BMP:

- Educate the public about hazardous materials in the home and consequences of improper use or disposal.
- Identify and promote the use of nonhazardous alternatives.
- Identify proper storage and disposal methods.
- Promote HHW reuse and recycling.
- Promote participation in local HHW collection programs.
- Distribute posters, handouts, and educational efforts aimed at local schools.
- Use public service announcements on local television, radio, and newspapers.
- Try utility bill inserts.
- Make video or slide presentations to community organizations.
- Develop a "speakers bureau" made up of local environmental professionals and recycling experts.

LIMITATIONS. This BMP maybe limited to areas with convenient access to hazardous waste disposal facilities and recycling facilities because of the cost associated with transport. This BMP can be a high-cost option compared to other source controls. There are significant liability issues involved with the collection, handling, and disposal of household hazardous waste.

EXAMPLES OF EFFECTIVE PROGRAMS. There are a number of communities, using a variety of approaches, with established and effective HHW collection programs. Seattle/King County, Washington, uses a mobile collection program that was initiated in 1989. One of the oldest (1988) and most convenient permanent collection centers in the country is in San Francisco, California. The Regional Water Quality Control Plant in Palo Alto, California, hosts a periodic program on the first Saturday morning of each month.

USED OIL RECYCLING

Used oil recycling is a responsible alternative to improper disposal practices, such as dumping oil in the sanitary sewer or storm drain system, applying oil to roads for dust control, placing used oil and filters in the trash for landfill disposal, or simply pouring used oil on the ground.

APPROACH. The following approaches may be effective for used oil recycling:

- Integrate efforts with a municipal solid waste program that likely has already been established.

- Set up a municipal collection center funded by the city.
- Contract out the collection and hauling of used oil to a private hauler or recycler.
- Use the automobile service industry (for example, service stations and fast-oil-change businesses) for the collection of used oil.
- Work with automotive parts supply stores and their parking lots, where consumers often change their automotive fluids improperly.

COST CONSIDERATIONS. A collection facility or curbside collection may result in significant costs. Using commercial locations (such as automobile service stations and fast-oil-change businesses) as collection points eliminates hauling and recycling costs for a municipality.

Staffing costs are minimal when using commercial locations as collection points; staffing costs are higher if the city performs collection services.

REGULATORY CONSIDERATIONS. Some states have enacted legislation requiring the state agency to pay a recycling incentive to curbside collection programs and certified used oil collection centers. Municipalities must comply with all applicable state and federal regulations regarding storage, handling, and transport of petroleum products. The municipality may be required to have a used oil recycling element within its integrated waste management plan.

ADMINISTRATIVE AND STAFFING CONSIDERATIONS. Staffing requirements are minimal if collection and recycling are contracted out to a used oil hauler or recycler or required at commercial locations.

PUBLIC EDUCATION AND PARTICIPATION CONSIDERATIONS. Create procedures for collection, such as collection locations and schedules, acceptable containers, and maximum amounts accepted. Promote public participation through the use of posters, handouts, brochures, and announcements in the print and broadcast media; provide a list of commercial recyclers. Also, develop incentive programs (such as a return deposit) for commercial locations and used oil haulers or recyclers.

LIMITATIONS. The availability of reliable, licensed used oil haulers and recyclers may be limited. The program requires frequent public education/notification messages. The used oil/hazardous waste separation requirement under federal law may also be a limitation. Meeting zoning, fire, and health and safety laws associated with collecting used oil may not be possible at all locations.

EXAMPLES OF EFFECTIVE PROGRAMS. There are numerous locations throughout the country that accept used oil. As is the case with HHW, communities have used different methods to collect and recycle used oil. Examples of effective programs include the permanent Household Hazardous

Waste Collection Facility in San Francisco, California, and curbside collection programs in Sacramento and Palo Alto, California. Statewide programs include California Integrated Waste Management Boards's program in which businesses that accept used oil from the public are listed as "Certified Oil Collection Centers" and receive payment for those collections. Another nationally recognized used oil program is Project R.O.S.E. (Recycled Oil Saves Energy), operated by the University of Alabama and funded by the Science, Technology, and Energy Division of the Alabama Department of Economic and Community Affairs.

VEHICLE SPILL CONTROL

Methods for preventing or reducing the discharge of pollutants to stormwater from vehicle leaks and spills include reducing the chance for spills by preventive maintenance, stopping the source of spills, containing and cleaning up spills, properly disposing of spill materials, and training employees. This BMP covers only prevention and cleanup of spills from vehicles; it does not contain information on underground storage tanks.

APPROACH. Vehicles will leak and spill fluids. The key is to reduce the frequency and severity of leaks and spills and, when they do occur, prevent or reduce the environmental effects. The following approaches to vehicle spill control may be effective:

- Perform fluid removal and changes inside or under cover on paved surfaces.
- Properly store hazardous materials and waste.
- Have spill cleanup supplies readily available.
- Clean up spills and leaks immediately.
- Use dry cleanup methods.
- Prepare a written contingency plan between local agencies that outlines responsibilities for major spills from tanker trucks.

COST CONSIDERATIONS. The prevention of leaks and spills is inexpensive. Treatment and disposal of contaminated soil or water can be expensive. Keep ample supplies of spill control and cleanup materials at municipal facilities, near storage and maintenance areas. Also, update spill cleanup materials as changes occur in the types of chemicals stored on site.

REGULATORY CONSIDERATIONS. The federal government and most states have specific laws concerning oil, oil filters, and batteries.

ADMINISTRATIVE AND STAFFING CONSIDERATIONS. This BMP has no significant administrative or staffing requirements. Training is crucial to reduce the frequency, severity, and effects of leaks and spills.

PUBLIC EDUCATION AND PARTICIPATION CONSIDERATIONS. The following considerations may be useful for vehicle spill control:

- Encourage the general public to regularly inspect and maintain their vehicles.
- Educate car repair shops to identify leaks and advise the vehicle owner.

LIMITATIONS. For larger spills, a private spill cleanup company or hazardous materials team may be necessary.

EXAMPLES OF EFFECTIVE PROGRAMS. Typically, more advanced HHW and used oil collection programs include educational outreach to do-it-yourselfers (DIYs) who do their own oil changes. Using state grant funds, San Mateo County, California, developed a DIY program that links used oil disposal with stormwater quality. The Santa Clara Valley and Alameda countywide stormwater programs joined the Palo Alto Regional Water Quality Control Plant to develop and distribute "Keeping It All in Tune," a car care brochure in multiple languages for DIYs.

ABOVEGROUND TANK SPILL CONTROL

Prevention or reduction of the discharge of pollutants to stormwater from aboveground storage tanks can be done by installing safeguards against accidental releases, installing secondary containment, conducting regular inspections, and training employees in standard operating procedures and spill cleanup techniques.

APPROACH. Integrate efforts with existing aboveground petroleum storage tank programs through the local fire and health departments and with area business emergency response plans through the city, county, or fire district. Use engineering safeguards to reduce the chance for spills. Perform regular maintenance. Also, keep ample supplies of spill control and cleanup materials at municipal facilities. Update spill cleanup materials as changes occur in the types of chemicals stored on site.

COST CONSIDERATIONS. Costs will vary depending on the size of the facility and the necessary controls.

REGULATORY CONSIDERATIONS. Consider requiring smaller secondary containment areas (less than 60 m [200 sq ft]) to be connected to the sanitary sewer and prohibiting any hard connections to the storm drain.

ADMINISTRATIVE AND STAFFING CONSIDERATIONS. This BMP has no significant administrative or staffing requirements. However, employees should be trained in spill prevention and cleanup.

LIMITATIONS. For larger spills, a private spill cleanup company or hazardous materials team may be necessary.

ILLEGAL DUMPING CONTROL

The use of measures to detect, correct, and enforce against the illegal dumping of pollutants in gutters and streets and into the storm drain system and creeks can have a significant effect on stormwater quality. This BMP includes controls of indirect sources (for example, overwatering after pesticide use on a lawn) and direct sources.

The remedial focus of this BMP contrasts with the preventive focus of the material disposal and recycling BMPs in this chapter. Illegal discharges through physical connections are addressed in the section on illicit connection BMPs.

APPROACH. Public awareness is the key to this BMP. Train municipal employees and educate the general public to recognize and report illegal dumping. Deputize municipal staff with the authority to write environmental tickets. Establish a system for tracking incidents. Consider indirect sources to be as important and typical, if not more so, than direct sources.

COST CONSIDERATIONS. The primary cost is for staff time. The cost depends on how aggressively a program is implemented. Additionally, a database is useful for defining and tracking the magnitude of the problem.

REGULATORY CONSIDERATIONS. Municipal codes should include sections prohibiting the discharge of soil, debris, refuse, hazardous wastes, and other pollutants to the storm sewer system. State laws may authorize the confiscation or impoundment of vehicles involved in illegal dumping.

ADMINISTRATIVE AND STAFFING CONSIDERATIONS. This BMP requires technical staff to detect and investigate illegal dumping violations and coordinate public education. Legal staff members are required to pursue prosecution. The training of technical staff in identifying and documenting illegal dumping incidents is also required.

PUBLIC EDUCATION AND PARTICIPATION CONSIDERATIONS. Educate the public about antidumping ordinances (fold into existing household hazardous waste program). Awareness of the issue accomplishes two things: the receiver of the information understands the issue and, therefore, is unlikely to cause a problem, and the public's awareness often helps detect other violations.

LIMITATIONS. The elimination of illegal dumping depends on the availability, convenience, and cost of alternative means of disposal. Some communities encourage homeowners and commercial establishments to rake yard or green waste into the street for pickup. For the most part, the general public does not know when they are using the storm drain system, which leads to a large volume of indirect discharges.

STREET CLEANING

Some reduction in the discharge of pollutants to stormwater from street surfaces can be accomplished by conducting street cleaning on a regular basis.

APPROACH. The following approaches may be effective in implementing and maintaining the street-cleaning BMP:

- Prioritize cleaning to use the most sophisticated sweepers, at the highest frequency, and in areas with the highest pollutant loading.
- Optimize cleaning frequency based on interevent times (for example, the dry period between storms).
- Increase sweeping frequency just before the rainy season.
- Keep in mind that proper maintenance and operation of sweepers greatly increases their efficiency.
- Keep accurate operation logs to track program.

COST CONSIDERATIONS. Any street-cleaning program for water quality improvement requires a significant capital and O & M budget. Sweeper costs range from $65 000 to $120 000 per machine, depending on the type. There is a definite cost–benefit relationship between increased sweeping efficiency/frequency and pollutant removal that a municipality should understand before making significant changes to its existing street-sweeping program.

REGULATORY CONSIDERATIONS. Densely populated areas or heavily used streets may require parking regulations to clear streets for cleaning.

ADMINISTRATIVE AND STAFFING CONSIDERATIONS. The following considerations may apply to the street-cleaning BMP:

- Sweeper operators and maintenance, supervisory, and administrative personnel are required.
- Traffic control officers may be required to enforce parking restrictions.

- Skillful design of cleaning routes is required for a program to be productive.
- Arrangements must be made for disposal of collected wastes.
- Operators must be trained in proper sweeper operation.

PUBLIC EDUCATION AND PARTICIPATION CONSIDERATIONS. The general public should be educated about the need to obey parking restrictions and participate by using litter receptacles to reduce street litter.

LIMITATIONS. The following limitations may apply to this BMP:

- No currently available conventional sweeper is effective at removing oil and grease.
- Mechanical sweepers are not effective at removing fine sediment.
- Parked cars are the primary obstacles to effective street sweeping.
- Effectiveness may also be limited by street condition, traffic congestion, presence of construction projects, climatic conditions, and condition of curbs.

EXAMPLES OF EFFECTIVE PROGRAMS. In San Francisco, California, 90% of the streets are swept at least once per week, and some sections are swept two to three times per week. San Francisco is also converting as much of its fleet as possible to vacuum sweepers. The effects of these actions are less clear in separate sewer areas, but there are examples of other effective programs. These include the city of Beverly Hills, California, which sweeps all streets in the commercial district six times per week and all streets in the residential area at least once per week.

CATCH BASIN CLEANING

Catch basin and stormwater inlet maintenance should be done on a regular basis to remove pollutants, reduce high pollutant concentrations during the first flush of storms, prevent clogging of the downstream conveyance system, and restore the catch basin's sediment-trapping capacity.

APPROACH. Catch basins should be cleaned regularly enough to reduce the possibility of sediment and pollutant loading from the flushing effect of stormwater inflow. A catch basin that becomes a source rather than a sink for sediments is not being cleaned frequently enough. Prioritize maintenance to clean catch basins and inlets in areas with the highest pollutant loading and in areas near sensitive water bodies. Keep accurate operation logs to track the program.

COST CONSIDERATIONS. A catch basin cleaning program requires a significant capital and O & M budget because of the typically large number of catch basins in any given area and the high cost of labor and specialized equipment to do the work. Except for small communities, with relatively few catch basins that may be cleaned manually, most municipalities will require mechanical cleaners such as eductors, vacuums, or bucket loaders.

REGULATORY CONSIDERATIONS. There are no regulatory requirements for this BMP. Municipal codes should include sections prohibiting the disposal of soil, debris, refuse, hazardous waste, and other pollutants into the storm sewer system.

ADMINISTRATIVE AND STAFFING CONSIDERATIONS. The following administrative and staffing considerations may apply to this BMP:

- Two-person teams are required to clean catch basins with vacuum trucks.
- Arrangements must be made for the proper disposal of collected wastes.
- Crews must be trained in proper maintenance, including recordkeeping, disposal, and safety procedures.

PUBLIC EDUCATION AND PARTICIPATION CONSIDERATIONS. For this BMP to be successful, educate contractors (cement, masonry, painting) and utility employees (telephone, cable, gas, and electric) about proper waste disposal.

LIMITATIONS. The metal content of the decant and solids cleaned from a catch basin should be periodically tested to determine if the decant violates limits for disposal to the wastewater treatment plant or if the solids would be classified as a hazardous waste.

VEGETATION CONTROLS

Vegetation control typically involves a combination of mechanical methods and careful application of chemicals (herbicides). Mechanical vegetation methods are discussed herein; vegetation control by herbicides is addressed in the housekeeping practices BMP described earlier in this chapter. Mechanical vegetation control includes leaving existing vegetation, cutting less frequently, hand cutting, planting low-maintenance vegetation, collecting and properly disposing of clippings and cuttings, and educating employees and the public.

APPROACH. The following are areas of concern:

- Steep slopes,
- Vegetated drainage channels,

- Creeks,
- Areas adjacent to catch basins, and
- Detention/retention basins.

These areas are of less concern to stormwater quality:

- Flat or relatively flat vegetated areas,
- Areas not adjacent to drainage structures, and
- Areas screened from drainage structures by vegetation.

COST CONSIDERATIONS. Additional costs will result from upgrading certain mowing equipment for bagging and hiring additional laborers involved in cutting by hand and picking up clippings.

REGULATORY CONSIDERATIONS. Local municipal antidumping ordinances can be used to ensure that when vegetation is controlled by cutting or removal the waste is disposed of properly. Grading ordinances often prescribe controls and limits on exposure of soil after removal of vegetation. In an effort to meet solid waste reduction goals, many municipalities require or encourage composting yard waste instead of landfill disposal.

ADMINISTRATIVE AND STAFFING CONSIDERATIONS. Additional labor may be needed to hand cut and pick up clippings from areas where mechanical cutting and collection are not practical.

PUBLIC EDUCATION AND PARTICIPATION CONSIDERATIONS. Educate the public about careful use of or alternatives to herbicides, proper disposal of clippings and cuttings, and the effect of erosion from exposed soil.

LIMITATIONS. The public may not find existing, natural, or low-maintenance vegetation as attractive or desirable as ornamental or higher maintenance vegetation in some situations.

STORM DRAIN FLUSHING

A storm drain is "flushed" with water to suspend and remove deposited materials. Flushing is particularly beneficial for storm drain pipes with grades too flat to be self-cleansing. Flushing helps ensure pipes convey design flow and removes pollutants from the storm drain system.

APPROACH. Locate reaches of storm drain with deposit problems and develop a flushing schedule that keeps the pipe clear of excessive buildup. Also, consider flushing portions of the storm drain system upstream of the problem area, including gutters and streets, as a preventive measure and an opportunity to remove more pollutants (assuming the collection system is sufficient).

COST CONSIDERATIONS. Unless flushing is done to a dry or wet detention area or the sanitary sewer, the collection of liquid and sediments may be costly in terms of pollutant removal benefits.

The following equipment may be required:

- Water source (water tank truck, fire hydrant),
- Sediment collector (evacuator/vacuum truck, dredge),
- Inflatable bladders to block flow from exiting pipe, or
- Sediment and turbidity containment and treatment equipment, if flushing to an open channel.

ADMINISTRATIVE AND STAFFING CONSIDERATIONS. With storm drain flushing, the following considerations should be noted:

- Two-person teams are needed for routine sediment removal and flush water collection.
- Equipment operators are required.

PUBLIC EDUCATION AND PARTICIPATION CONSIDERATIONS. If large-scale flushing activities are undertaken, local residents should be informed in advance. The public should be educated about the purpose of storm drains and the problems created by illegal dumping.

LIMITATIONS. These limitations may apply:

- Flushing is most effective in small-diameter pipes.
- The availability of sufficient water to do the job must be ensured.
- Personnel may have difficulty finding a downstream area to collect sediments.
- Flushing requires liquid/sediment disposal.
- The disposal of flushed effluent to the sanitary sewer may be prohibited in some areas because of inflow capacity and water quality concerns of the local wastewater treatment plant.

ROADWAY AND BRIDGE MAINTENANCE

Methods to prevent or reduce the discharge of pollutants to stormwater from roadway and bridge maintenance include paving as little area as possible, designing bridges to collect and convey stormwater, using measures to prevent runon and runoff, properly disposing of maintenance wastes, and training employees and subcontractors.

APPROACH. Address stormwater pollution from roadway and bridge maintenance on a site-specific basis. The disposition and subsequent magnitude of pollutants found in road and bridge runoff is variable and affected by climate, surrounding land use, roadway or bridge design, traffic volume, and frequency and severity of accidental spills.

COST CONSIDERATIONS. This BMP is typically low in cost. Additionally, keep ample supplies of drip pans or absorbent materials on site.

REGULATORY CONSIDERATIONS. Consider requiring new bridges to incorporate treatment control BMPs to the design. For more information on treatment control BMPs, see Chapter 5 of this publication.

ADMINISTRATIVE AND STAFFING CONSIDERATIONS. This BMP has no significant administrative or staffing requirements. Inspect the activities of employees and subcontractors to ensure that measures to reduce the stormwater effects of roadway and bridge maintenance are being followed. Require engineering staff or consulting firms to address stormwater quality in new bridge designs or existing bridge retrofits.

LIMITATIONS. There are no significant limitations to this BMP.

EXAMPLES OF EFFECTIVE PROGRAMS. The county of Alameda, California, recently wrapped a 56-year old bridge with tarpaulins so that workers could remove lead paint without allowing toxics to fall into the local estuary or blow away in the wind.

DETENTION AND INFILTRATION DEVICE MAINTENANCE

Proper maintenance and sediment removal are required on both a routine and corrective basis to promote effective stormwater pollutant removal efficiencies for wet and dry detention pond and infiltration devices.

APPROACH. These approaches may be beneficial for a detention and infiltration device maintenance BMP:

- Remove silt after sufficient accumulation.
- Periodically clean accumulated sediment and silt from pretreatment inlets.

- Infiltration device silt removal should occur when the infiltration rate drops below 13 mm (0.5 in.) per hour.
- Removal of accumulated paper, trash, and debris should occur at least every 6 months or as needed to prevent clogging of control devices.
- Vegetation growth should not be allowed to exceed approximately 4.5 m (18 in.) in height.
- Mow the slopes periodically and check for clogging, erosion, and tree growth on the embankment.
- Corrective maintenance may require more frequent attention.

COST CONSIDERATIONS. Frequent sediment removal is labor intensive and costly. Transport and disposal costs for waste material will vary proportionately with the volume of material. Disposal costs can be high if sediments have high levels of toxics. Other cost considerations are vehicles, dump trucks, bulldozers, backhoes, excavators, mowers, weed trimmers, sickles, machetes, shovels, rakes, and personal protective equipment (goggles, dust masks, coveralls, boots, and, gloves).

ADMINISTRATIVE AND STAFFING CONSIDERATIONS. Two-person teams are needed for routine silt removal and excavation. A program manager is needed to track maintenance activities and provide field assistance. A staff team is needed for corrective maintenance activities. Training in appropriate excavation and maintenance procedures and proper waste disposal procedures is needed.

REGULATORY CONSIDERATIONS. Permits may be required by the U.S. Army Corps of Engineers, U.S. Fish and Wildlife Service, or state parks and wildlife agencies.

PUBLIC EDUCATION AND PARTICIPATION CONSIDERATIONS. It may be useful to create a public education campaign to explain the function of wet and dry detention pond and infiltration devices and their operational requirements for proper effectiveness. Also, encourage the public to report wet and dry detention pond and infiltration devices needing maintenance.

LIMITATIONS. Dredging sediments in a wet detention pond produces slurried waste that often exceeds the limits for acceptability used by many landfills. Frequent sediment removal is labor and cost intensive. Care should be taken when using monitoring wells so as not to provide a conduit to the unsaturated zone under basins, which might lead to groundwater contamination.

EXAMPLES OF EFFECTIVE PROGRAMS. Because of flat terrain and few water courses, the city of Fresno, California, built almost 100 stormwater retention/recharge basins to dispose of surface runoff and recharge an aquifer. Monitoring confirmed that a variety of organic and inorganic contaminants generated in the catchments are removed by sorption within the top 4 cm of sediment in the recharge basins, making these contaminants available for removal and disposal through routine maintenance. Monitoring results also show that contaminants have not degraded groundwater quality beneath the basins.

STORM CHANNEL AND CREEK MAINTENANCE

Reduction of pollutant levels in stormwater can be achieved by regularly removing illegally dumped items and material from storm drainage channels and creeks. Channel characteristics can be modified to enhance pollutant removal and hydraulic capacity.

APPROACH. The following approaches may be beneficial to storm channel and creek maintenance:

- Identify illegal dumping "hot" spots; conduct regular inspection and cleanup of hot spots and other storm drainage areas where illegal dumping and disposal occurs.
- Post "No Littering" signs with a phone number for reporting a dumping in progress.
- Adopt and enforce substantial penalties for illegal dumping and disposal.
- Modify storm channel characteristics to improve channel hydraulics, increase pollutant removals, and enhance channel and creek aesthetic and habitat value.
- Maintain accurate logs to evaluate materials removed and improvements made.
- Establish buffer zones along creeks.

COST CONSIDERATIONS. The following cost considerations may apply to this BMP:

- The purchase and installation of signs,
- The cost of vehicle(s) to haul illegally disposed items and material to landfills,
- The rental of heavy equipment to remove larger items (for example, car bodies) from channels,
- The purchase of landfill space to dispose of illegally dumped items and material, and
- Capital and maintenance costs for channel modifications.

REGULATORY CONSIDERATIONS. Regulatory considerations include the adoption of substantial penalties for illegal dumping and disposal.

ADMINISTRATIVE AND STAFFING CONSIDERATIONS. Larger municipalities should commit at least one full-time staff person; smaller municipalities should commit at least one part-time staff person. Additional staff can be added as needed. Staff will need training in channel maintenance and use of heavy equipment and training in the identification and handling of hazardous materials and wastes.

PUBLIC EDUCATION AND PARTICIPATION CONSIDERATIONS. The storm channel and creek maintenance BMP may require the following:

- Education on the need for proper disposal of refuse;
- Notification of penalties for illegal dumping and disposal; and
- Promotion of volunteer services to create litter collection groups (such as Adopt-a-Stream).

LIMITATIONS. Cleanup activities may create a slight disturbance for local aquatic species. Access to items and material on private property may be limited. Tradeoffs may exist between channel hydraulics and water quality habitat. Worker and public safety may be at risk in high-crime areas.

ILLICIT CONNECTION PREVENTION

Preventing unwarranted physical connections to the storm drain system from sanitary sewers and floor drains through regulation, regular inspection, testing, and education can remove a significant source of stormwater pollution.

APPROACH. The following steps are components of this BMP:

- Ensure that existing municipal building and plumbing codes prohibit physical connections of nonstormwater discharges to the storm drain system.
- Require visual inspection of new developments or redevelopments during the development phase.
- Develop documentation and recordkeeping protocols to track inspections and catalog the storm drain system.
- Use techniques such as zinc chloride smoke testing, fluorometric dye testing, and television camera inspection to verify physical connections.

COST CONSIDERATIONS. Zinc chloride smoke testing, fluorometric dye testing, and television camera inspection can be cost considerations. Also, there may be additional labor and equipment costs for verification of plumbing connections.

REGULATORY CONSIDERATIONS. Ensure that existing building and plumbing codes prohibit physical connection to the storm drain system of nonstormwater discharges, and establish penalties for such action. Implement procedures to inspect and verify that new construction development does not interface with the storm drain system.

ADMINISTRATIVE AND STAFFING CONSIDERATIONS. Building and plumbing inspectors must verify and document that inappropriate discharges are not allowed through connections to the storm drain system. Zinc chloride smoke testing, fluorometric dye testing, or television inspection may be necessary for the verification of illicit connections. Additional staff time is required for verification and documentation of proper connections to the storm drain system.

PUBLIC EDUCATION AND PARTICIPATION CONSIDERATIONS. Consider a community awareness program (using various media), targeting appropriate audiences (homeowners, businesses, and contractors) to warn against improper connections to the storm drain system and encourage public reporting of illegal connections through a community hot line telephone number. Notify community and local fire departments before testing with zinc chloride smoke testing and fluorometric dye testing in targeted areas.

LIMITATIONS. The following limitations may be applicable:

- Proper connections may be verified on date of inspection but could be altered afterwards by illicit connections.
- The cost for inspection equipment can be high.
- The removal of an illicit connection from the storm drain system to the sanitary sewer may require the approval of the local sewer authority.
- Improper physical connections to the storm drain system can occur in a number of ways, such as the overflow of crossconnects from sanitary sewers and floor drains from businesses such as auto shops and restaurants.

ILLICIT CONNECTION—DETECTION AND REMOVAL

Control procedures for the detection and removal of illegal connections from the storm drain conveyance system should be implemented to reduce or prevent unauthorized discharges to receiving waters. Procedures include field screening, follow-up testing, and complaint investigation.

APPROACH. The approach can involve any or all of the following:

- Field screening program;
- Sampling program, including beach sampling during dry weather;
- Fluorometric dye testing (suspected source testing);
- Zinc chloride smoke testing (suspected source testing);
- Television camera inspection;
- Physical inspection testing;
- Citizens' complaints on a community hotline to report suspected illegal connections; and
- Correction by plugging, disconnecting, or otherwise eliminating the discharge route.

COST CONSIDERATIONS. Considerations for this BMP can include program initialization costs for procuring necessary equipment and training. Explore the possibility of equipment sharing with municipal wastewater treatment departments.

Keep in mind that illegal connection location techniques can be labor and equipment intensive.

The following equipment may be necessary:

- Personal protective equipment such as hardhats, boots, plastic gloves, and coveralls;
- Sampling containers and equipment;
- U.S. EPA-recommended stormwater test kits;
- Fluorometric dye and fluorometer (optional);
- Zinc chloride smoke and dispersal fans;
- Pipeline television camera with videocassette recorder;
- Self-contained breathing apparatus;
- Oxygen/toxic/combustible gas detection meters;
- Vehicle(s); and
- Aboveground communication.

ADMINISTRATIVE AND STAFFING CONSIDERATIONS. Administrative and staffing considerations include the following items:

- Two-person teams are required for field screening and sampling activities.
- Larger staff teams (at least one additional member) are required for fluorometric dye and zinc chloride smoke testing or television camera inspection and physical inspection with confined space entry.
- Staffing a community hotline telephone number may be necessary.

- Program coordination is needed for emergencies and recordkeeping.
- Health and safety training required by the Occupational Safety and Health Administration (OSHA) (29 CFR 1910.120), with annual refresher training, may be needed.
- Confined space entry training (federal OSHA 29 CFR 1910.146) may be needed as required by OSHA.
- Procedural training (field screening, sampling, dye or smoke testing, television or other inspection, source training, and corrective and mitigative techniques) of staff may be necessary.

PUBLIC EDUCATION AND PARTICIPATION CONSIDERATIONS. Encourage public reporting of improper waste disposal or evidence of illicit connections. Community notification, including notifying the local fire department, is required in targeted areas for zinc chloride smoke testing.

LIMITATIONS. Local ordinances must specify access rights to private property to allow for field screening and testing along storm drain system right-of-ways. Additionally, local ordinances may require that the source be suspected for guaranteed rights of entry to conduct verification testing.

EXAMPLES OF EFFECTIVE PROGRAMS. The Fort Worth, Texas, Drainage Water Pollution Control Program is action-oriented and designed to monitor and correct nonstormwater discharges in urban storm drain systems using innovative screening, tracing, and corrective techniques. Staff requirements are one full-time staff member equivalent per year (three part-time staff members are used from other programs) for a total labor cost of $30 000. Equipment, supplies, and services range from $20 000 to $50 000 per year, depending on the level of activity and co-utilization of city services. Costs could be higher for initial start-up and equipment purchases. If another program were unable to use existing city staff, additional costs would be expected.

LEAKING SANITARY SEWER CONTROL

Control procedures should be implemented for identifying, repairing, and remediating infiltration, inflow, and wet weather overflows from sanitary sewers to the storm drain conveyance system. Procedures include field screening, follow-up testing, and compliance investigation.

APPROACH. The approaches listed below may be useful for sanitary sewer control:

- Identify dry weather infiltration and inflow first. Wet weather overflow connections are difficult to locate.
- Locate wet weather overflows and leaking sanitary sewers using conventional source identification techniques, including
  - — Field screening program (including field analytical testing),
  - — Fluorometric dye testing,
  - — Zinc chloride smoke testing,
  - — Television camera inspection,
  - — Nessler reagent test kits for ammonia detection, and
  - — Citizens' hot line for reporting of wet weather sanitary overflows.

COST CONSIDERATIONS. Cost considerations include the following:

- There may be program costs for procuring necessary equipment and training.
- Departmental cooperation is recommended for sharing or borrowing staff resources and equipment from municipal wastewater treatment departments. Infiltration, inflow, and wet weather overflows from sanitary sewers can be labor and equipment intensive to locate.
- The following equipment may be needed:
  - — Personal protective equipment (such as hardhats, boots, plastic gloves, and coveralls);
  - — Sampling containers/equipment;
  - — Stormwater test kits (recommended by U.S. EPA);
  - — Zinc chloride smoke and dispersal fans;
  - — Fluorometric dye and fluorometer (optional);
  - — Pipeline television camera with videocassette recorder;
  - — Self-contained breathing apparatus;
  - — Oxygen/toxic/combustible gas detection meters;
  - — Vehicle(s), and
  - — Aboveground communication.

ADMINISTRATIVE AND STAFFING CONSIDERATIONS. Two-person teams are needed to perform field screening and associated sampling. Larger teams (at least one additional member) are required for fluorometric dye testing, zinc chloride smoke testing, television camera inspection, and physical inspection with confined space entry. Program coordination is required for handling emergencies and recordkeeping.

Health and safety training (required by OSHA, 29 CFR 1910.129), with annual refresher training, is needed. Confined space entry training (federal OSHA, 29 CFR 1910.146) may be needed. Also, procedural training (field screening, sampling, dye/smoke testing, television or other inspection, source training, and corrective/mitigative techniques) of staff may be needed.

PUBLIC EDUCATION AND PARTICIPATION CONSIDERATIONS. Consider a public awareness program through local media to identify the problem of sanitary infiltration, inflow, and wet weather overflows to the storm sewer system.

Establish a community response hotline for reporting observed sanitary infiltration or leaks and wet weather sanitary overflows to the storm sewer system. Finally, remember that public notification, including notifying the local fire department, is required for fluorometric dye or zinc chloride smoke testing in targeted areas.

LIMITATIONS. The local ordinance must specify access rights to private property to allow for field screening and testing along storm drain system right-of-ways. Also, local ordinances may require that the source be suspected for guaranteed rights of entry to conduct verification testing.

EXAMPLES OF EFFECTIVE PROGRAMS. The city of Stockton, California, Municipal Stormwater Discharge Management Program contains a comprehensive program element created to prevent, detect, and eliminate illegal connections to storm sewers.

The city of Fort Worth, Texas, Drainage Water Pollution Control Program is an action-oriented program designed for corrective measures using innovative biotoxicity testing. Staffing requirements are one full-time person per year for a total labor cost of $30 000. Equipment, supplies, and services costs approximately $50 000 per year.

REFERENCES

Camp Dresser & McKee *et al.* (1993) *California Storm Water Best Management Practices Handbooks.* Public Works Agency, County of Alameda, Calif.

City of Stockton (1993) NPDES Storm Water Permit Application. Part II, Calif.

County of San Bernardino (1993) NPDES Drainage Area Management Program. San Bernardino County Flood Control District—Santa Ana Basin Area, Calif.

Livingston, E.H., *et al.* (1988) *The Florida Development Manual, A Guide to Sound Land and Water Management.* Dep. Environ. Regulation, Tallahassee, Fla.

SUGGESTED READINGS

Alameda County Urban Runoff Clean Water Program (1994) Street Sweeping/Storm Inlet Modification Literature Review. Hayward, Calif.

Alameda County Urban Runoff Clean Water Program (1992) Public Information/Participation Plan. Hayward, Calif.

American Public Works Association (1978) Street Cleaning Practice. Chicago, Ill.

California Stormwater Quality Task Force (1994) *Stormwater Resource Guide, A Listing of the Materials Available in California Relating to Stormwater and Watershed Management.* Public Information/Public Participation Subcommittee, Sacramento, Calif.

Camp Dresser & McKee (1992) Municipal Stormwater Discharge Management Program—City of Stockton, California. Walnut Creek, Calif.

City of Austin (1989) *Environmental Criteria Manual, Design Guidelines for Water Quality Control.* Tex.

City of Austin (1989) *Environmental Criteria Manual: Land Development Code.* Tex.

City of Olympia (1994) Impervious Surface Reduction Study: Technical and Policy Analysis—Final Report. Public Works Dep., Wash.

City of San Jose (1994) Riparian Corridor Policy Study. Calif.

City of Seattle (1989) Water Quality Best Management Practices Manuals. Wash.

Florida Department of Environmental Regulation (1988) *Florida Development Manual: A Guide to Sound Land and Water Management.* Tallahassee, Fla.

Gartner, Lee & Associates (1983) Toronto Area Watershed Management Strategy Study. Technical Report #1—Humber River and Tributary Dry Weather Outfall Study, Ont. Ministry Environ., Toronto, Ont., Can.

Glanton, T., *et al.* (1991) The Illicit Connection—EPA Stormwater Regulations Field Screening Program and the City of Houston's Successful Screening Program. Paper presented at Tex. Water Pollut. Control Assoc. Conf., Tex.

Glanton, T., *et al.* (1992) The Illicit Connection—Is it the Problem? *Water Environ. Technol.,* **4,** 9, 63.

Lahor, M.M., *et al.* (1991) Cross-Connection Investigations For Stormwater Permit Applications. Paper presented at 64th Annu. Conf. Water Pollut. Control Fed., Toronto, Ont., Can.

Lazaro, T.R. (1990) *Urban Hydrology: A Multidisciplinary Perspective.* Technomic Publishing Company, Inc., Lancaster, Pa.

Minnesota Pollution Control Agency (1989) Protecting Water Quality in Urban Areas: Best Management Practices for Minnesota. St. Paul, Minn.

Moore, R., *et al.* (1992) Applications of EPA's Stormwater Discharge Regulations in Boston, MA. Paper presented at Water Environ. Fed. Collection Systems Symp., New Orleans, La.

Office of Planning and Research (1986) The California Environmental Quality Act. State Calif. Governor's Office, Office of Permit Assistance, Calif.

Office of Planning and Research (1990) Planning, Zoning, and Development Laws. State Calif. Governor's Office, Office Plann. Res., Calif.

Ontario Ministry of Environment and Energy (1994) *Stormwater Management Practices Planning and Design Manual.* Toronto, Ont., Can.

Ortolano, L. (1984) *Environmental Planning and Decision Making.* John Wiley & Sons, New York, N.Y.

Pelletier, G. J., and Determan, T.A. (1988) *Urban Storm Drain Inventory Inner Gray Harbor.* Wash. State Dep. Ecol., Water Qual. Investigations Section, Olympia.

Phillips, N.J., and Lewis, E.T. (1993) Site Planning from a Watershed Perspective. In *National Conference on Urban Runoff Management; Enhancing Urban Watershed Management at the Local, County, and State Levels.* U.S. EPA, Chicago, Ill.

Pisano, W. (1989) Feasibility Study For Stormwater Treatment For Belmont Street Drain. Environmental Design & Planning, Inc., Dep. Public Works, City of Worcester, Mass.

Pitt, R., and McLean, J. (1986) Toronto Area Watershed Management Strategy Study, Humber River Pilot Watershed Project. Final Rep., Ont. Ministry Environ., Toronto, Ont., Can.

Pitt, R., *et al.* (1993) *Investigation of Inappropriate Pollutant Entries into Storm Drainage Systems—A User's Guide.* EPA-600/R-92-238, Office Res. Dev., Washington, D.C.

Regional Water Quality Control Plant (1995) Clean Bay Plan—1995. City of Palo Alto, Calif.

San Francisco Department of Public Works (1992) Best Management Practices Public Education Plan. Calif.

San Francisco Household Hazardous Waste Collection Facility (1992) Fourth Year–1991, Annual Report, City and County of San Francisco, Calif.

Santa Clara Valley Nonpoint Source Pollution Control Program (1990) Public Information/Participation Plan. Santa Jose, Calif.

Santa Clara Valley Nonpoint Source Pollution Control Program (1992) Best Management Practices for Automotive-Related Industries. San Jose, Calif.

Santa Clara Valley Nonpoint Source Pollution Control Program (1992) Best Management Practices for Industrial Stormwater Pollution Control. San Jose, Calif.

Sartor, J.D., and Gaboury, D.R. (1984) Street Sweeping as a Water Pollution Control Measure: Lessons Learned Over the Past Ten Years. *Sci. Total Environ.,* **33**, 171.

Schmidt, S.D., and Spencer, D.R. (1986) The Magnitude of Improper Waste Discharges in an Urban Stormwater System. *J. Water Pollut. Control Fed.,* **58**, 744.

Schumacher, J.W., and Grimes, R.F. (1992) A Model Public Education Process for Stormwater Management. *Public Works*, 55.

State of Florida (1988) *The Florida Development Manual, A Guide to Sound Land and Water Management.* Dep. Environ. Regulation, Fla.

Terrene Institute (1991) Handle with Care: Your Guide to Prevention Water Guide to Preventing Water Pollution. Washington, D.C.

Urban Drainage and Flood Control District (1992) *Urban Storm Drainage Criteria Manual.* Volume 3—Best Management Practices—Stormwater Quality, Denver, Co.

Uribe and Associates (1990) Best Management Practices Program for Pollution Prevention. City and County of San Francisco, Oakland, Calif.

U.S. Environmental Protection Agency (1979) *Dry Weather Deposition and Flushing for Combined Sewer Overflow, Pollution Control.* EPA-600/2-79-133, Washington, D.C.

U.S. Environmental Protection Agency (1987) *Guide to Nonpoint Source Pollution Control.* Washington, D.C.

U.S. Environmental Protection Agency (1988) *Used Oil Recycling.* EPA-530/SW-89-006, Washington, D.C.

U.S. Environmental Protection Agency (1989) *How to Set Up a Local Program To Recycle Used Oil.* EPA-530/SW-89-039A, Washington, D.C.

U.S. Environmental Protection Agency (1990) *Collecting Household Hazardous Wastes at Wastewater Treatment Plants: Case Studies.* EPA-430/09-90-016, Washington, D.C.

U.S. Environmental Protection Agency (1990) *National Pollutant Discharge Elimination System Permit Application Regulations for Storm Water Discharges.* Final Rule, 40 CFR Parts 122, 123, and 124, *Fed. Regist.* 55 (222).

U.S. Environmental Protection Agency (1991) *Proposed Guidance Specifying Management Measures for Sources of Nonpoint Pollution in Coastal Waters.* Washington, D.C.

U.S. Environmental Protection Agency (1992) *Designing an Effective Communication Program: A Blueprint for Success.* Region V, Washington, D.C.

U.S. Environmental Protection Agency (1992) *Manual of Practice Identification of Illicit Connections.* EPA-833/R-90-100, Washington, D.C.

U.S. Environmental Protection Agency (1992) *Storm Drainage Systems—A User's Guide.* EPA-600/R-92-238, Office Res. Dev., Washington D.C.

U.S. Environmental Protection Agency (1992) *Stormwater Management for Construction Activities: Developing Pollution Prevention Plans and Best Management Practices.* EPA-833/R-92-005, Washington, D.C.

U.S. Environmental Protection Agency (1992) *Stormwater Management for Industrial Activities: Developing Pollution Prevention Plans and Best Management Practices.* EPA-832/R-92-006, Washington, D.C.

U.S. Environmental Protection Agency (1993) *Investigation of Inappropriate Pollutant Entries into Storm Drainage Systems—A User's Guide.* EPA-600/R-92-238, Washington, D.C.

U.S. Environmental Protection Agency (1993) Urban Stream Restoration. *Proc. Natl. Conf. Urban Runoff Manage.,* Chicago, Ill.

U.S. Geological Survey (1995) Potential for Chemical Transport Beneath a Storm-Runoff Recharge (Retention) Basin for an Industrial Catchment in

Fresno, California. Water Resour. Investigations Rep. 93-4140, prepared in cooperation with Fresno Metropolitan Flood Control Dist., Calif.
Washington State Department of Ecology (1991) Vactor Truck Operations and Disposal Practices. Olympia, Wash.
Washington State Department of Ecology (1992) *Stormwater Management Manual for the Puget Sound Basin (The Technical Manual).* Volume IV—Urban Land Use BMPs. Olympia, Wash.
Washington State Department of Ecology (1993) Contaminants in Vactor Truck Wastes. Olympia, Wash.
Water Environment Federation (1995) *Controlling Vehicle Service Facility Discharges in Wastewater. How to Develop and Administer a Source Control Program.* Alexandria, Va.
Water Environment Research Foundation (1996) *Residential and Commercial Source Control Assessment.* Alexandria, Va.

Chapter 5
Selection and Design of Passive Treatment Controls

The selection and successful design of selected passive treatment controls often called structural best management practices (BMPs) for stormwater quality enhancement is the cornerstone of stormwater management in newly developing and redeveloping urban areas. It is also possible sometimes to retrofit BMPs in already developed parts of a municipality; however, the costs can be high.

Overall, structural BMPs are most applicable to developing and redeveloping areas. The cost effectiveness of each control has to be considered and measured against the actual environmental benefits realized. One price of urban society is the indelible mark it leaves on the ecology of urban areas. Through the use of structural BMPs, these effects can be, in part, mitigated as new urban centers develop and the old ones redevelop or expand. It is unlikely, however, that all effects of urbanization can be eliminated.

This chapter will address how several known structural BMPs can be evaluated for use at any given site and, after being selected, how each can be sized and configured. The control practices described are those that have a known performance track record. Other practices and types of controls are evolving and may eventually prove to be superior to those described herein. However, this manual is limited to established state of practice at the time of publication. New and evolving practices will be the topic of future updates. Structural BMPs require commitment of resources for initial construction and continuing operation and maintenance. Despite the development of technology mandates that new concepts be tried and tested, the bottom line is that whatever controls are used, the designer, the client, and often the regulator need to have sufficient confidence in their performance before attempting to use them. As a result, the testing of newer concepts is sometimes better left to government agencies or risk takers in industry, who can provide the new design technologies to the stormwater professionals.

This chapter will cover the following topics:

- Stormwater quality hydrology;
- Selection of treatment facilities (that is, structural BMPs);
- Operation and maintenance needs of treatment facilities;
- Swales and filter strips;
- Stormwater infiltration, including minimized connected impervious area;
- Extended detention;
- Retention (wet) ponds;
- Constructed wetlands;
- Media filtration; and
- Oil and water separators.

The above topics can be better understood when the reader has a thorough understanding of the municipal system and its relationship to stormwater quality and quantity management infrastructure. Issues such as community needs and fiscal strength, local priorities and preferences, regulatory demands, other infrastructure needs and fiscal demands, and many other's are part of the equation. Also, this chapter is most useful if the reader has become familiar with the topics discussed in Chapters 1 through 4.

Ideally, structural BMPs should be a part of the treatment train discussed in Chapter 4—namely, source control BMPs. Good housekeeping measures need to be practiced to ensure adequate performance and longevity of structural facilities. For example, if erosion control during construction is not being rigorously practiced within the catchment being served by the structural control facility, the facility will probably be rendered inoperative in a short time. For example, a detention basin, a retention pond, or a constructed wetland will fill with sediment, and infiltration devices will fail. Thus, without implementing source erosion controls, the investment in the structural facilities will be lost, and expensive rehabilitative maintenance or reconstruction of these facilities will then be needed to return them to a working condition.

It is best that the practices described in Chapter 5 be selected through a comprehensive planning process. This could involve systemwide simulation of rainfall and runoff processes, preferably using a continuous model. Some of the selection of practices actually may be based on monitoring, bioassessments, or the understanding of effects on the receiving systems. Chapters 1, 2, and 3 address all of those topics and form the foundation and an introduction to all topics covered in this chapter.

HYDROLOGY FOR THE MANAGEMENT OF STORMWATER QUALITY

In 1990, the U.S. Environmental Protection Agency (U.S. EPA) promulgated National Pollution Discharge Elimination System regulations regarding the permitting of stormwater discharges from municipal storm sewer systems. These rules require the municipalities to reduce the pollutants in urban runoff to the maximum extent practicable (MEP). The definition of MEP for the control of stormwater pollutant discharges has focused primarily on the application of economically achievable management practices. Because stormwater runoff rates and volumes vary from storm to storm, the statistical probabilities of runoff events and their management have to be considered in the development of practices to meet the MEP goal. It is paramount that the hydrology of urban runoff be examined within this context.

The type and size of storm runoff events to use for the design of runoff treatment systems need to be examined. "Treatment systems" are those measures that are often referred to in the literature as "BMPs." Among these are swales; buffer strips; infiltration basins and percolation trenches; extended detention basins; "wet" ponds that retain some or all of one event's runoff until it is displaced by the runoff from a subsequent event; media filters; and a variety of other devices and facilities. Guidelines for the design of these types of facilities can be found in this chapter and elsewhere (Livingston *et al.*, 1988; Roesner *et al.*, 1989; Schueler, 1987; Urbonas and Roesner [Eds.], 1986; and Urbonas and Stahre, 1993).

LONG-TERM RAINFALL CHARACTERISTICS. Hydrologists typically look at the infrequent events: either large storms for drainage and flood protection or drought periods for water supply development. But what characteristics are representative of the storms that produce most rainfall on a long-term basis?

Figure 5.1 presents the cumulative probability distribution of daily precipitation data for 40 years at Orlando, Florida, and Cincinnati, Ohio. These data have been screened to include only precipitation events 2.5 mm (0.1 in.) or greater in Cincinnati and 1.5 mm (0.06 in.) or greater in Orlando. Cumulative occurrence probabilities were computed for values ranging from 2.5 to 51 mm (0.1 to 2.0 in.).

Examination of Figure 5.1 reveals most of the daily values to be less than 25 mm (1 in.) in total depths. In Orlando, which averages 1 270 mm (52 in.) of rainfall per year, 90% of these events produce less than 36 mm (1.4 in.) of rainfall. In Cincinnati, which has 1 016 mm (40 in.) per year of precipitation, 90% of the events produce less than 20 mm (0.8 in.) of rainfall. By contrast,

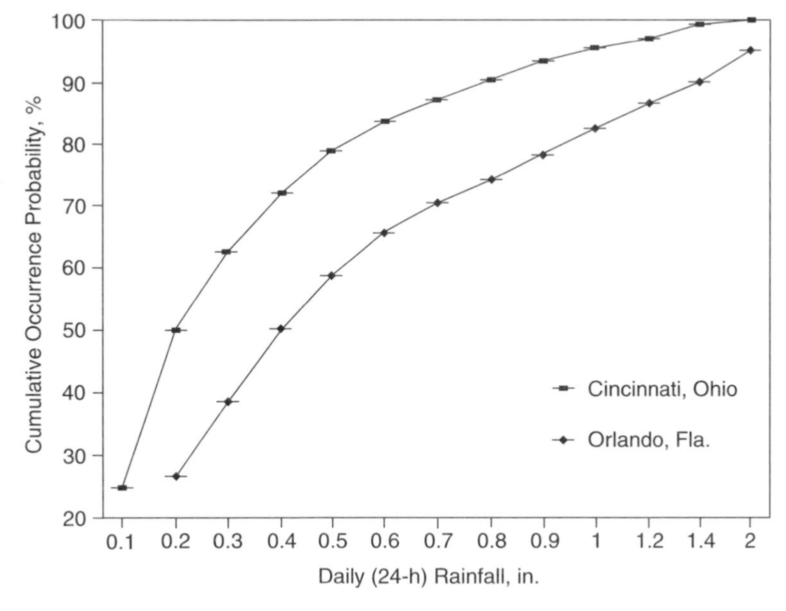

Figure 5.1 Cumulative probability distribution of daily precipitation for two cities in the U.S. (in. × 25.4 = mm) (Roesner *et al.*, 1991).

the 2-year, 24-hour storm produces precipitation of 127 mm (5.0 in.) in Orlando and 74 mm (2.9 in.) in Cincinnati. This suggests that capturing and treating runoff from "smaller" storms should capture a large percentage of the runoff events and runoff volume that occur from the urban landscape. Also, a water quality facility capable of capturing these smaller storms would also capture the "first flush" portion of the larger, infrequently occurring runoff events.

CAPTURE OF STORMWATER RUNOFF. To illustrate the terms "smaller" and "most" discussed earlier, long-term simulations of runoff were examined for six U.S. cities by Roesner *et al.* (1991) using the Storage, Treatment, Overflow, Runoff Model (STORM). The six cities were Butte, Montana; Chattanooga, Tennessee; Cincinnati, Ohio; Detroit, Michigan; San Francisco, California; and Tucson, Arizona. The Storage, Treatment, Overflow, Runoff Model is a simplified hydrologic model that translates a time series of hourly rainfall to runoff then routes the runoff through detention storage.

Hourly precipitation records of 40 to 60 years were processed by Roesner *et al.* (1991) for a variety of detention basin sizes for the six cities. These simulations were performed using the characteristics of the most typically occurring urban developments found in each city. Table 5.1 lists the average annual rainfall and the area-weighted runoff coefficient at each of the study watersheds. Runoff capture efficiencies of detention basins were tested using an outflow discharge rate that emptied or drained the design storage volume in 24 hours. This drawdown time was based on field study findings by Grizzard *et al.* (1986) in the Washington, D.C., area. They determined that a detention basin had to be designed to empty out a volume equal to the average runoff event's volume in no less than 24 hours to be an effective stormwater quality enhancement facility. The findings by Roesner *et al.* (1991) are illustrated in Figure 5.2.

One way to define a cost-effective basin size is to represent it as that which is located on the "knee of the curve" for capture efficiency. This "knee" is evident on the six curves in Figure 5.2. Urbonas *et al.* (1990) have

Table 5.1 Hydrologic parameters used at six study watersheds (Roesner *et al.*, 1991).

City	Average annual rainfall, in. (mm)	Runoff coefficient of study watershed
Butte, Montana	14.6 (371)	0.44
Chattanooga, Tennessee	29.5 (749)	0.63
Cincinnati, Ohio	39.9 (1 013)	0.50
Detroit, Michigan	35.0 (889)	0.47
San Francisco, California	19.3 (490)	0.65
Tucson, Arizona	11.6 (295)	0.50

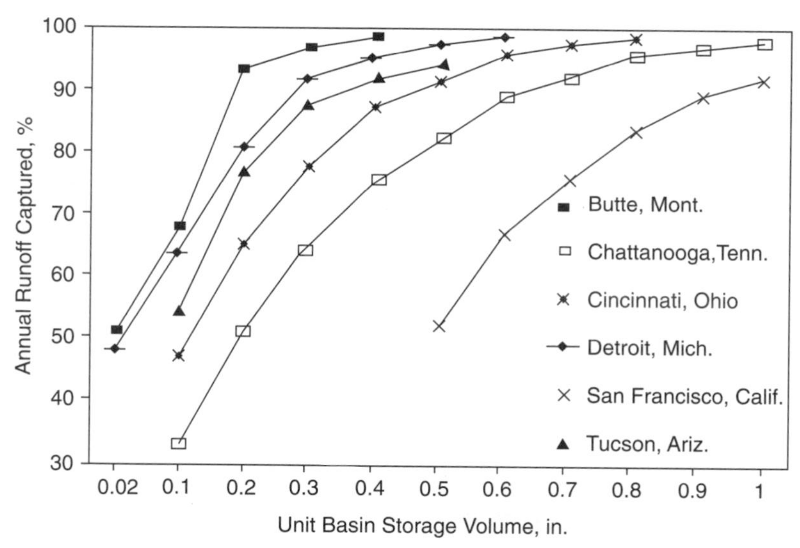

Figure 5.2 Runoff capture rates versus unit storage volume at six study sites (Roesner *et al.*, 1991).

defined this "knee" as the "optimized" capture volume and reported on a sensitivity study they performed relative to this volume for the Denver, Colorado, area. Later, Urbonas and Stahre (1993) redefined this "knee" as the "maximized" volume because it is the point at which rapidly diminishing returns in the number of runoff events captured begin to occur. For each of the six study watersheds previously described, the maximized storage volume values are listed in Table 5.2.

The sensitivity investigation by Urbonas *et al.* (1990) also estimated the average annual stormwater removal rates of total suspended sediments using the maximized volume as the surcharge storage above a permanent pool of a retention pond. Estimates of total suspended sediment removals were performed using the procedure reported by Driscoll (1983). Similarly, the runoff

Table 5.2 Maximized unit storage volume at six study watersheds (Roesner *et al.*, 1991).

City	Maximized storage volume[a]	
	in. (mm)	ac-ft/ac (m^3/ha)
Butte, Montana	0.25 (6.4)	0.021 (63.5)
Chattanooga, Tennessee	0.50 (12.7)	0.042 (127)
Cincinnati, Ohio	0.40 (10.2)	0.033 (102)
Detroit, Michigan	0.30 (7.6)	0.025 (76.2)
San Francisco, California	0.80 (20.3)	0.067 (203)
Tucson, Arizona	0.30 (7.6)	0.025 (76.2)

[a] Based on the ratio of runoff volume captured from all storms.

Table 5.3 Sensitivity of the best management practice capture volume in Denver, Colorado (Urbonas *et al.*, 1990).

Capture volume to maximized volume ratio	Annual runoff volume captured, %	Number of storms completely captured	Average annual total suspended sediments removed, %
0.7	75	27	86
1.0	85	30	88
2.0	94	33	90

capture and total suspended sediment removal efficiencies were estimated for capture volumes equal to 70 and 200% of the maximized volume. These findings are summarized in Table 5.3.

Review of Table 5.3 shows that doubling of the maximized capture volume results in a very small increase in the total annual runoff volume captured and an insignificant increase in the average annual removal of total suspended sediments. When 70% of the maximized volume is used, only a moderate decrease occurs in the volume of runoff captured and an insignificant decrease in the annual total suspended sediment load removed. Based on these findings, the Denver, Colorado, municipal area adopted an 80th percentile runoff event (that is, 95% of the maximized event) as the basis for the sizing of stormwater quality BMPs. This 80th percentile runoff event is now considered by the municipalities in this semiarid region of the U.S. as cost effective for stormwater quality management and is viewed as the design event that achieves MEP definition under the Clean Water Act.

Although the MEP event is not clearly defined by the regulations, insight to the appropriate MEP design event can be gained by performing an analysis of local long-term hourly rainfall data similar to those reported in Tables 5.1 though 5.3. These analyses form a basis for making a rational decision in defining sizing criteria for various BMPs. As an example, the maximized unit runoff volume for a watershed in Denver, Colorado, with a runoff coefficient $C = 0.5$, is 7 mm (0.28 watershed in.), or 70 m^3/ha (0.023 ac-ft/ac). This compares well with the maximized storage volumes listed in Table 5.2 for Butte, Montana, and Tucson, Arizona—namely, the two semiarid communities on that list.

As can be seen from Figure 5.2 and Tables 5.2 and 5.3, most runoff-producing events occur from the predominant population of smaller storms, namely, less than 13 to 25 mm (0.5 to 1.0 in.) of precipitation. To be effective, stormwater quality management should be designed based on these smaller events. As a result, detention facilities, wetland basins, infiltration facilities, media filters, and possibly swales need to be sized to accommodate runoff volumes and flows from such storm events to maximize pollution control benefits in a cost-effective manner.

AN APPROACH FOR ESTIMATING STORMWATER QUALITY CAPTURE VOLUME. Estimating a Maximized Water Quality Capture Volume. Whenever local resources permit, the stormwater quality capture volume may best be found using continuous hydrologic simulation and local long-term hourly (or lesser time increment) precipitation records (see Chapter 3). However, it is possible to obtain a first-order estimate of the needed capture volume using simplified procedures that target the most typically occurring population of runoff events.

Figure 5.3 contains a map of the contiguous 48 states of the U.S. with the mean annual runoff-producing rainfall depths superimposed (Driscoll *et al.,* 1989). These mean depths are based on a 6-hour interevent time to define a new storm event and a minimum depth of 2.5 mm (0.10 in.) of precipitation for a storm to produce incipient runoff. After an extensive analysis of a number of long-term precipitation records from different meteorological regions of the U.S., Guo and Urbonas (1995) found simple regression equations to relate the mean precipitation depths in Figure 5.3 to "maximized" water quality runoff capture volumes (that is, the knee of the cumulative probability curve).

The analytical procedure was based on a simple transformation of each storm's volume of precipitation to a runoff volume using a coefficient of runoff. To help with this transformation, a third-order regression equation, Equation 5.1 (Urbonas *et al.,* 1990), was derived using data from more than 60 urban watersheds (U.S. EPA, 1983). Because the data were collected nationwide over a 2-year period, Equation 5.1 should have broad applicability in the U.S. for smaller storm events.

$$C = 0.858i^3 - 0.78i^2 + 0.774i + 0.04 \quad (5.1)$$

Where

C = runoff coefficient, and
i = watershed imperviousness ratio; namely, percent total imperviousness divided by 100.

Equation 5.2 relates mean precipitation depth taken from Figure 5.3 to the "maximized" detention volume. The coefficients listed in Table 5.4 are based on an analysis of long-term data from seven precipitation gauging sites located in different meteorological regions of the U.S. The correlation of determination coefficient, r^2, has a range of 0.80 to 0.97, which implies a strong level of reliability.

$$P_0 = (a \cdot C) \cdot P_6 \quad (5.2)$$

Where

P_0 = maximized detention volume determined using either the event capture ratio or the volume capture ratio as its basis, watershed in. (mm);

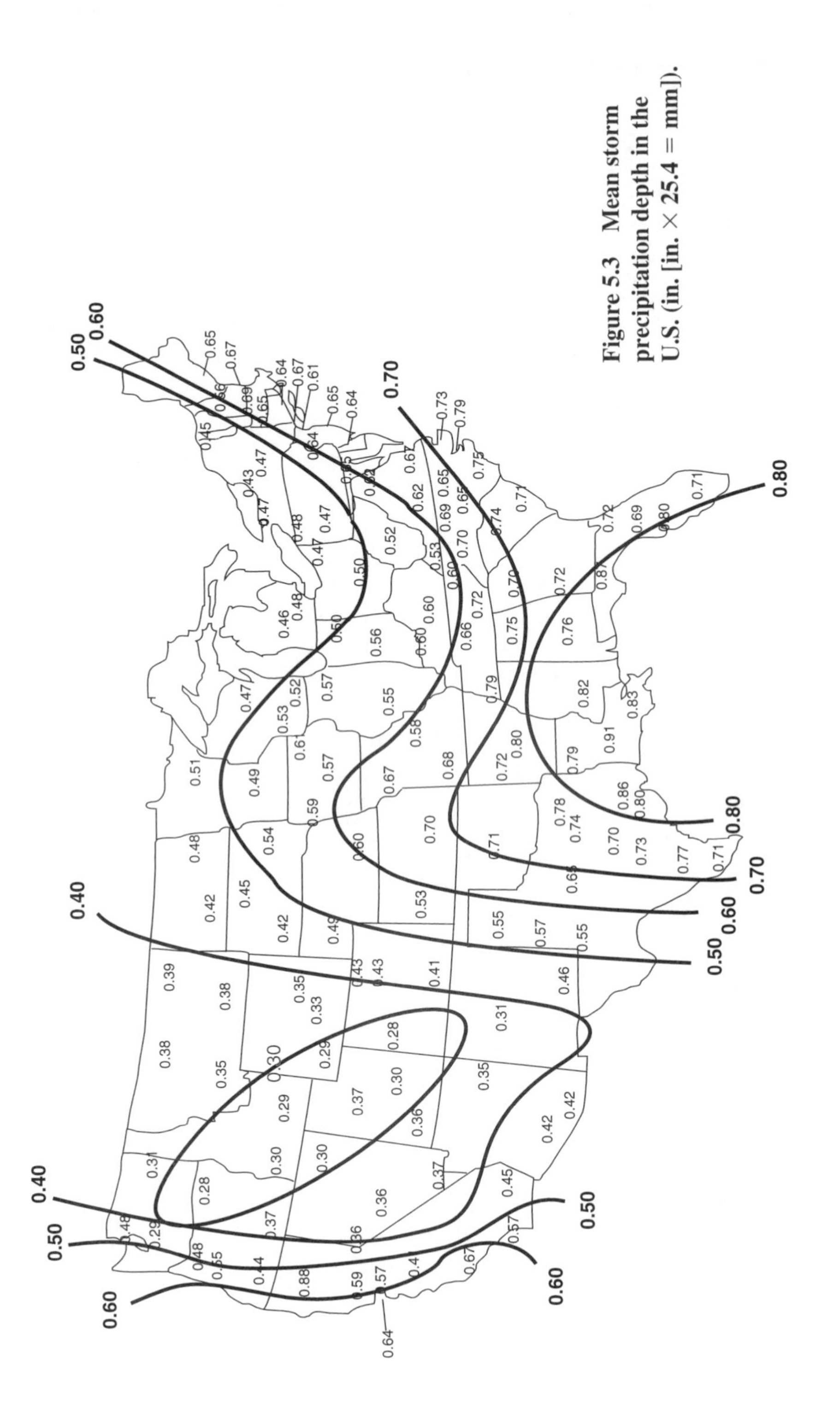

Figure 5.3 Mean storm precipitation depth in the U.S. (in. [in. × 25.4 = mm]).

Table 5.4 Values of coefficient *a* in Equation 5.2 for finding the maximized detention storage volume (Guo and Urbonas, 1995).[a]

		Drain time of capture volume		
		12 hours	24 hours	48 hours
Event capture ratio	a =	1.109	1.299	1.545
	r^2 =	0.97	0.91	0.85
Volume capture ratio	a =	1.312	1.582	1.963
	r^2 =	0.80	0.93	0.85

[a] Approximately 85th percentile runoff event (range 82 to 88%).

a = regression constant from least-squares analysis;
C = watershed runoff coefficient; and
P_6 = mean storm precipitation volume, watershed in. (mm).

Table 5.4 lists the maximized detention volume/mean precipitation ratios based on either the ratio of the total number of storm runoff events captured or the fraction of the total stormwater runoff volume from a catchment. These can be used to estimate the annual average maximized detention volume at any given site. All that is needed is the watershed's runoff coefficient and its mean annual precipitation.

The actual size of the runoff event to target for water quality enhancement should be based on the evaluation of local hydrology and water quality needs. However, examination of Table 5.3 indicates that the use of larger detention volumes does not significantly improve the average annual removal of total suspended sediments or other settleable constituents. It is likely that an extended detention volume equal to a volume between the runoff from a mean precipitation event taken from Figure 5.3 and the maximized event obtained using Equation 5.2 will provide the optimum-sized and most cost-effective BMP facility. A BMP sized to capture such a volume will also capture the leading edge (that is, first flush) of the runoff hydrograph resulting from larger storms.

Runoff volumes that exceed the design detention volume either bypass the facility or receive less efficient treatment than do the smaller volume storms and have only a minimal net effect on the detention basin's performance. If, however, the design volume is larger and has an outlet to drain it in the same amount of time as the smaller basin, the smallest runoff events will be detained only for a brief interval by the larger outlet. Analysis of long-term precipitation records in the U.S. shows that small events always seem to have the greatest preponderance. As a result, oversizing the detention can cause the most frequent runoff events to receive less treatment than provided by properly designed smaller basins.

Example of a Water Quality Capture Volume Estimate. It is desired to estimate the maximized storage volume for a 223-ha (550-ac) watershed that has 40% of its area covered by impervious surfaces. Assume that this site is located in Houston, Texas (that is, the largest storm region of the U.S.). The detention basin needs to be sized and designed to drain its water quality capture volume in no fewer than 24 hours. Substituting a value of 0.40 (that is 40/100) for the variable I in Equation 5.1 yields a runoff coefficient C = 0.30. Using Figure 5.3 we find the mean storm precipitation depth in Houston: $P6$ = 20.3 mm (0.8 in.). From Table 5.4 we find the coefficient a = 1.299 for the 24-hour drain time. Thus, the maximized detention volume is calculated as follows:

$$P_0 = (1.299 \cdot C) = 0.39 \text{ in. } (0.026 \text{ ac-ft/ac})$$

$$P_0 = 7.9 \text{ mm } (79 \text{ m}^3/\text{ha})$$

The volume of an extended detention basin for this 223-ha (550-ac) watershed needs to be 17 600 m^3 (14.3 ac-ft). It is recommended that this volume be increased by at least 20% to account for the loss in volume from sediment accumulation. The final design then can show a total volume for the basin of 21 200 m^3 (17.2 ac-ft) with an outlet designed to empty out the bottom 17 600 m^3 (14.3 ac-ft) of this volume in approximately 24 hours.

SELECTION OF TREATMENT CONTROL BEST MANAGEMENT PRACTICES

As discussed in Chapter 4, the selection of particular stormwater quality controls (that is, BMPs) is often decided by considerations other than technical issues. These issues, among others, include

- Federal, state, and local regulations;
- Real and perceived receiving water problems;
- Beneficial uses of receiving waters to be protected;
- The cost of the BMPs being considered;
- Subjective and sometimes arbitrary acceptance by the regulators or community groups; and
- Watershed studies.

While reduction of pollutants in stormwater discharges to the MEP is the statutory requirement of the stormwater regulations, the real goal has to be the reduction of effects to urban stormwater runoff on the receiving water. The cost of the BMP is always a major consideration. Recognize that no single BMP is as effective as a "train" (that is series) of practices and controls. The selection and design of structural BMPs is the topic of Chapter 5.

POINTS TO CONSIDER WHEN SELECTING BEST MANAGEMENT PRACTICES. There are several general points that need to be considered whenever selecting and designing any treatment control. The following discussion highlights some of these points.

Source Control. As a first step in the treatment train, source controls should be considered before the more expensive treatment controls are selected. These good housekeeping practices can help to reduce the amount of pollutants coming into contact with stormwater and being transported to receiving waters. Because the structural treatment controls will not remove all pollutants, the use of source controls will supplement, and thereby increase, the efficiency of the total stormwater quality management system.

Local Climate. Many treatment controls rely on the "wet" state where vegetation, biological processes, or the presence of a permanent pool can enhance pollutant removals. In arid and semiarid areas such as the southwest, such treatment controls are not practical unless water losses resulting from evapotranspiration are replaced in part or in total by municipal or irrigation supplies. The state of practice today has emerged from the observations of facilities in climates where the rainfall season is coincident with the growth of vegetation and evaporation losses of pond surfaces. The designer needs to make certain that controls being selected are compatible with local climatology and availability of water to keep these controls functional. For example, a wetland basin in a semiarid climate may not have sufficient water to keep the wetland species alive or healthy unless supplemental irrigation water is provided, which may make this choice economically impractical.

Design Storm Size. The use of design storms for the sizing of water quality controls are not the same as those selected for the design of facilities to control runoff rates for urban drainage. For example, the effect on a receiving water by the pollutant washoff of a 25-year storm is inconsequential compared to the potential hydraulic damage. Of concern for water quality control are the small, frequent events, smaller than the 1-year storm, that carry the vast majority of runoff and pollutants. Stormwater quality hydrology is discussed earlier in this chapter.

Soil Erosion. The success of any BMP control facility is influenced by how rigorously soil erosion protection is being practiced in the tributary watershed. In arid and semiarid climates, native vegetation can be sparse, allowing for greater erosion during storms than in more moist climates. But, irrigated lawns in semiarid climates tend to reduce excessive erosion after the land becomes urbanized. Regardless of climate, higher-than-normal sediment loads will affect the performance and maintenance requirements of treatment controls.

Stormwater Pollutant Characteristics. Potential pollutants can be either in particulate or dissolved form. Some BMPs will only remove particulates. Other BMPs, such as wet ponds, are purported to remove dissolved constituents and particulates. However, the data confirming this premise are limited and sometimes contradictory. Whenever specific pollutants are of particular concern, the designer needs to select treatment controls that are most likely to best remove the constituents of greatest concern.

Multiple Uses. Consider the opportunities for integrating stormwater treatment needs with other community objectives. These include active and passive recreation, enhancing or protecting wildlife habitat, flood control, and groundwater recharge.

Maintenance. None of the treatment controls described herein requires active operation of equipment such as mechanical or chemical systems. Nonetheless, they need to be maintained to operate as designed. The operation and maintenance needs of each facility have to be addressed because their life-cycle costs should be considered.

Physical and Environmental Factors. The most typical physical and environmental factors that need to be considered in the selection of treatment controls are

- Slope—most BMPs are sensitive to the local terrain slope. Certain BMPs, such as swales and infiltration basins, cannot be used in steep terrain, while others, such as detention basins and filters, can be made to work on most reasonably sized land parcels.
- Area required—most BMPs require considerable land area. Some can be placed underground at considerable expense.
- Soil—infiltration systems must be located on porous soils and subsoils. Vegetation requires good soils, while wet ponds require impermeable soils or lining.
- Water availability—the BMPs that rely on vegetation or a permanent water pool for pollutant removal require water, which has to be available in adequate quantity during dry seasons.
- Aesthetics and safety—where visible or accessible to the public, aesthetics and safety are of concern.
- Hydraulic head—some BMPs need a drop in water elevation (for example, an extended detention outlet), which is a site topography issue.
- Environmental side effects—the design process should consider mos-

quito breeding, groundwater contamination, opportunities for aquatic and wildlife habitat, and active and passive recreation.

Regardless of the above factors, in all cases the treatment controls to be used must be compatible with existing drainage or flood control objectives.

To assist the user with the selection of treatment control BMPs in a given watershed, the decision tree shown in Figure 5.4 was developed. This decision tree will lead to treatment control BMPs that are potentially applicable

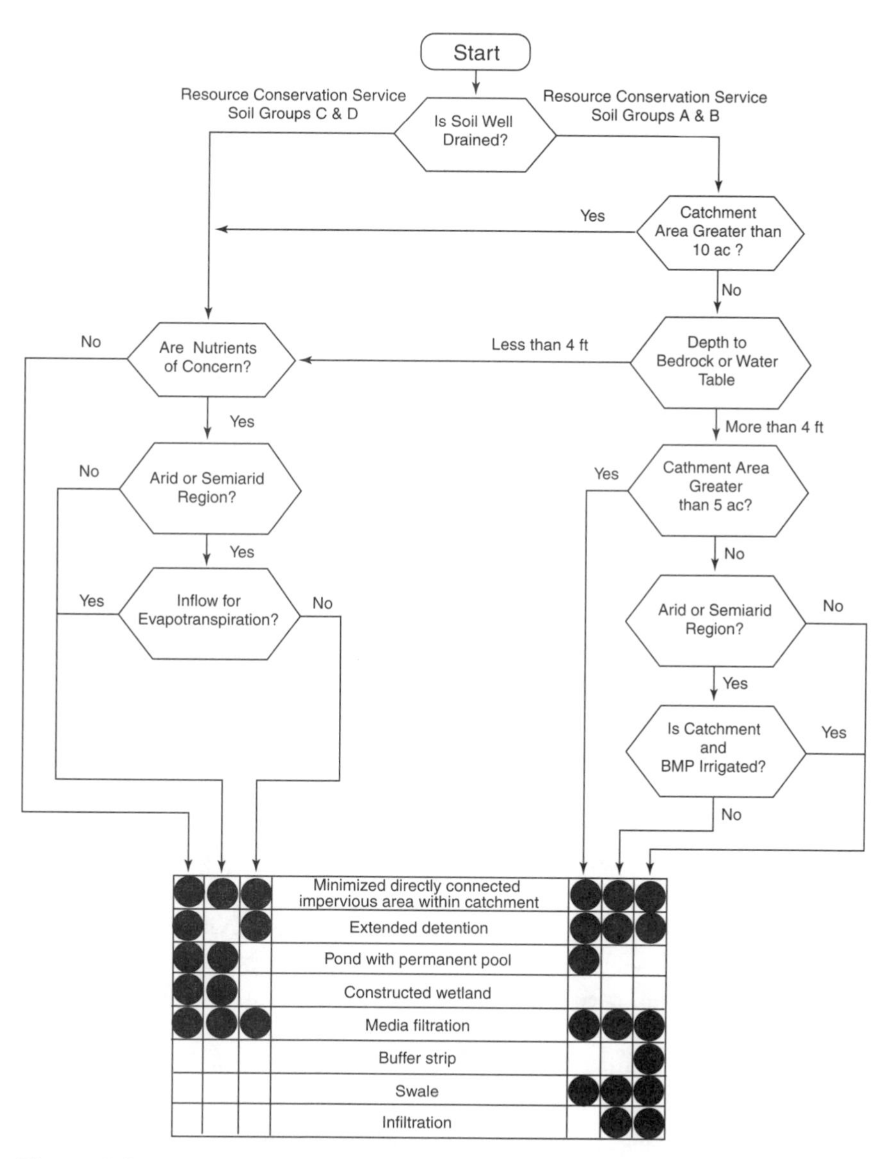

Figure 5.4 Decision tree for identifying potential treatment controls (ac × 0.404 = ha; ft × 0.304 8 = m).

to a given site/region. Using the site characteristics and the design guidance for the BMPs contained in this chapter, potentially applicable BMPs can be designed and their cost can be estimated. Table 5.5 can then be filled out for the watershed(s) or pollutant(s) of concern to narrow the focus to the most cost-effective BMPs appropriate for the specific site and purpose.

On-Site Versus Regional Controls. Some structural BMPs can be used either as stand-alone, on-site treatment controls or as part of regional controls for stormwater quality enhancement, while others can only be used on site. Swales, filter strips, infiltration and percolation, media filters, oil and water separators, and other controls are not applicable for large-scale, regional controls. Extended detention, wet retention ponds, and wetland basins can be used as regional controls serving large catchments and, if local conditions permit, can also be used as on-site controls. Because large numbers of on-site controls, sometimes exceeding several hundred or even several thousand, can eventually be installed within an urban watershed, it becomes difficult to reliably quantify their cumulative hydrologic effects on receiving waters. Water quality, however, can be improved by both regional and on-site controls, although the degree of improvement for the accumulated effect in numerous on-site controls is less predictable than with regional controls.

Large numbers of on-site controls complicate the quality assurance during design and construction because they are probably designed by a variety of individuals and are constructed by a variety of different contractors under varying degrees of quality control. Further, they may be maintained and operated in a variety of ways impossible to anticipate or to control.

A simple example illustrates what a municipality faces with on-site controls. If the average size of a new land development is 8 ha (20 ac) and only one type of a treatment BMP is installed for each new development, there will be 14 on-site controls per km^2 (32 on-site controls per sq mile). A city or county with more than 260 km^2 (100 sq miles) area would then have to keep track of 3 200 controls. It is not unusual for an on-site control to serve less than 4 ha (10 ac) or even 2 ha (5 ac), and the outlet diameters become small. To ensure maintenance for a large number of such controls, a municipality needs a large, properly trained staff to keep track of their condition. Merely mandating that these facilities be built when land development occurs is not enough.

A more logical approach is to use regional controls serving 32 to 240 ha (80 to 600 ac). This eliminates the uncertainty of large numbers of on-site controls. Large sites need multistage outlets to detain small events. As the watershed becomes larger and the outlets become bigger, runoff from small storms experience little "throttling." To compensate for this, multistage outlets can be used to release small runoff events in 12 to 24 hours and empty the total water quality capture volume in 24 to 48 hours.

Wiegand *et al.* (1989) estimated that regional controls are more cost effective because fewer controls are less expensive to build and maintain than

Table 5.5 Form for evaluating structural best management practices.

Worksheet for evaluation of treatment control best management practices								
Pollutant of concern	Best management practice	Area of application, ha	Annual pollutant removed, kg/a	Annual capital costs,[a] $/a	Annual operation and maintenance cost,[b] $/a	Annual administrative costs, $/a	Total annual costs, $/a	Removal cost[c] $/kg

[a] Annual capital costs based on a 20-year design period.
[b] Annual administration costs are best determined by a given community after a citywide program is established.
[c] Removal costs are in units of ($/a)/(kg/a) = $/kg.

a large number of on-site controls. Another benefit is that water quality control outlets are larger and are easier to design, build, operate, and maintain. Also, because regional controls are often maintained by some form of public body, they are more likely to receive ongoing maintenance.

Regional controls can provide treatment for existing and new developments and typically will capture all runoff from public streets, which is often missed by on-site controls. Because regional controls cover large land areas, this permits other compatible uses such as recreation, wildlife habitat, or aesthetic open space to occur within their boundaries.

The two major disadvantages of regional controls are that they require advanced planning and up-front financing. Lack of financing early in the watershed's land development process, before sufficient developer contributions are available, can preclude their timely installation. The use of on-site controls often is the only practical institutional and political alternative.

Overview of Specific Treatment Controls. Treatment controls, often called structural BMPs, can occupy much land area. They often require flat side slopes for access, ease of maintenance, and public safety. Although vertical sides can be used to reduce the needed land area, safety fencing may be needed to reduce the owner's potential liability, which then affects the facility's aesthetics.

The most frequently used treatment controls include swales, buffer strips, infiltration basins and trenches, and extended detention basins and ponds.

SWALES. Swales are shallow sideslope channels with a relatively mild longitudinal slope, typically grassed or vegetated. They are designed for slow velocities during small storms, allowing opportunity for infiltration along the swale bottom and for the trapping of sediment and organic biosolids in the vegetative cover.

FILTER STRIPS. Sometimes called buffer strips, filter strips perform in a manner similar to swales but are not channels. These are mildly sloping vegetated surfaces that are located off, and abut, an impervious surface area. They are designed to slow the velocity of the runoff from the impervious area, increasing the opportunities for infiltration and the trapping of pollutants.

INFILTRATION BASINS AND PERCOLATION TRENCHES. These treatment controls capture runoff generated by small storms from small catchments and, when they work, provide good stormwater treatment by transferring surface runoff to the groundwater regime. This filters out suspended pollutants and provides other treatment processes before water returns to the surface system.

DETENTION CONTROLS. These include extended detention basins (dry), which drain out completely between storm events, and retention ponds (wet), which "retain" some or all of the storm runoff from a given event within its permanent pool until the next storm occurs. Detention basins remove pollutants primarily through sedimentation of solids, while retention ponds remove pollutants through sedimentation and through physical and biochemical processes in the permanent pool during the dry weather periods that follow. Typically, retention ponds are larger than detention basins and can be more efficient than detention basins in the removal of many constituents found in stormwater.

WETLAND BASINS. Wetland basins can be efficient stormwater quality treatment controls. Because of the regulatory protection of wetlands in the U.S., whatever their origin may be, wetland basins for stormwater treatment must be artificially created. Even so, federal or state regulatory agencies can assume control of these artificially "constructed" wetlands, requiring the owners to obtain permits before they can perform needed maintenance. Failure to obtain such permits can become a problem for the owner, individual, or organization performing mechanical cleaning, excavation, or dredging operations within these control treatment facilities.

MEDIA FILTRATION. These controls include sand filters with presettlement to avoid clogging. Other media mixtures of peat, sand, and compost mix and geotextiles are used but with mixed results.

OIL AND WATER SEPARATORS. These controls are designed to remove petroleum compounds, grease, sand, and grit. These devices are designed to remove floatable debris.

Robustness of Design Technology. High robustness of design implies that, when all of the design parameters are correctly defined and quantified, the design has a high probability of performing as intended. In other words, the design technology is well established and has undergone the test of time. Low robustness implies that there are many uncertainties in how the design will perform over time. In the definition used here, it is assumed that the facility will be properly operated and maintained over time as we currently understand its maintenance needs (which are not factors in judging design robustness).

Table 5.6 summarizes the collective opinion of senior professional engineers involved in the field of stormwater quality management about the design robustness of various structural stormwater quality controls. Some of these opinions acknowledge the dependence on local site-specific conditions, such as climates, geology, soil types, and nature of constituents. In some cases, the hydraulic design governs how reliable the design will be and how adequately it will perform. In those cases, even when water quality enhancement performance is known to be excellent when the facility functions properly, its ability to provide water quality enhancement cannot be ensured when its hydraulic design robustness is low. The reverse is also true; namely, a fa-

Table 5.6 Robustness of best management practice design technology.[a]

Type	Hydraulic design	Removal of constituents in stormwater: Total suspended sediments and solids	Dissolved	General Performance
Swale	Moderate-high	Low-moderate	None-low	Low
Buffer strip	Low-moderate	Low-moderate	None-low	Low
Infiltration basin	Moderate-high[a]	High	Moderate-high	Moderate
Percolation trench	Low-moderate[a]	High	Moderate-high	Moderate-high
Extended detention	High	Moderate-high	None-low	Moderate-high
Wet retention pond	High	High	Low-moderate	Moderate-high
Wetland	Moderate-high	Moderate-high	Low-moderate	Low-high[b]
Media filter	Low-moderate	Moderate-high	None-low	Low-moderate
Oil separator	Low-moderate	Low	None-low	Low
Catch basin inserts	Technology unknown	NA[d]	NA	NA
Monolithic porous pavement[b]	Low-moderate	Moderate-high	Low-high[c]	Low-moderate
Modular porous pavement[b]	Moderate-high	Low-high	Low-high[c]	Low-high[c]

[a] Weakest design aspect, hydraulic or constituent removal, governs overall design robustness.
[b] Robustness is site-specific and maintenance dependent.
[c] Low-moderate whenever designed with an underdrain and not intended for infiltration and moderate-high when site conditions permit infiltration.
[d] Not applicable.

cility that has high hydraulic reliability and low water quality enhancement predictability will have a low combined design robustness. In other words, the weakest design link governs the overall reliability of the design.

MAINTENANCE OF TREATMENT CONTROLS

Regular inspection and maintenance of treatment controls (that is, BMPs) are a necessity if these controls are to consistently perform up to expectations. Sediment and debris removal from inlets and outlets are the most important requirements. The following discussion presents an overview of maintenance requirements. Further information may be found in *Design and Construction of Urban Stormwater Management Systems* (ASCE and WEF, 1992). Detailed maintenance requirements for specific treatment BMPs can be found among the references listed in this chapter.

INSPECTIONS. Inspections should be performed at regular intervals to ensure that the BMP is operating as designed. Annual inspection should be considered, with additional inspections following storm events. For the inspection following a major storm, the inspector should visit the site at the end of the specified drawdown period to ensure that any detention or infiltration device is draining properly. Some inspections can be arranged to coincide with scheduled maintenance visits to reduce site visits and ascertain that maintenance activities are performed satisfactorily. Check for accumulations of debris and sediment at the inlets and outlets, and check side slopes for signs of erosion, settlement, slope failure, or vehicular damage. Check emergent vegetation zones to ensure that water levels are appropriate for vegetative growth, that acceptable survival rates are being maintained, and that vegetative cover is above acceptable limits.

ROUTINE MAINTENANCE. Routine or preventive maintenance refers to procedures that are performed on a regular basis to keep the BMP aesthetic and in proper working order. Routine maintenance should include debris removal, silt and sediment removal, and clearing of vegetation around flow control devices to prevent clogging. It is expected that silt removal will have to be performed every 5 to 15 years, as needed.

Routine maintenance also includes the maintenance of a healthy vegetative cover. Dead turf or other unhealthy vegetative areas will need to be replaced after being discovered.

NONROUTINE MAINTENANCE. Nonroutine or corrective maintenance refers to any rehabilitative activity that is not performed on a regular basis. This includes flow control structure replacement or the major replacement and cleaning of aquatic vegetation.

Erosion and Structural Repair. Areas of erosion and slope failure should be repaired and reseeded (or sodded) as soon as possible. Eroded areas near the inlet or outlet may also need to be lined with riprap.

Any major damage to inlet, outlet, or other structures should be repaired immediately. Delay in such repairs can invite structural failure the next time the facility is in operation. When that occurs, it may require total reconstruction of the structure.

Sediment Removal and Disposal. Although considered by some to be routine maintenance, silt and other sediment removal, with few exceptions, is anything but routine. First, it does not occur annually in most treatment facilities. Second, when it is done, it is typically a project that requires mechanized equipment, careful survey, transport and disposal of removed materials, and the reestablishment of the original design grades and sections of the BMP.

Sediment may need to be removed on a regular schedule but rarely on an annual basis. The only exceptions are certain types of oil and grit separators, media filters, and infiltration systems. For most treatment systems, the exact schedule will depend on the annual total suspended sediment load being removed by the facility and the size of the area on which it is being deposited.

Several field observations reported accumulation rates of 6 to 13 mm per year in retention ponds. This rate of accumulation has also been reported to be 10 to 100 times greater whenever construction activities take place in the tributary watershed, especially when effective erosion control practices are not used.

Equation 5.3 (Urbonas, 1994) can be used to estimate the average depth of sediment accumulation within almost any facility that removes total suspended sediments from stormwater:

$$V_P = 1.45 \cdot 10^{-6} \cdot \frac{h \cdot T_{SS} \cdot f_r}{R} \tag{5.3}$$

Where

V_P	=	average annual depth of bottom sediment deposit, mm;
h	=	average annual runoff depth from the watershed, mm;
T_{SS}	=	average annual concentration of total suspended sediments in runoff, mg/L;
f_r	=	fraction of TSS retained in pond; and
R	=	(pond's surface area)/(tributary watershed area) ratio.

As an example, estimate the annual accumulation rate within a retention pond with a surface area of 0.53 ha (1.3 ac), a tributary catchment of 223 ha (550 ac) with a runoff coefficient $C = 0.28$, an annual runoff-producing precipitation of 352 mm (12.8 in.), and an average concentration of $T_{SS} = 400$ mg/L in the runoff, of which the retained fraction in the pond is $f_r = 0.80$. First, find the annual runoff depth from the watershed—namely, $h = 0.28 \times 352$ mm = 99 mm. Then, the average annual accumulation of sediment is

$$V_P = 1.45 \cdot 10^{-6} \cdot \frac{99 \cdot 400 \cdot 0.80}{0.002\ 4} = 19 \text{ mm } (0.75 \text{ in./yr})$$

If the pond's original design allowed for a total of 305 mm (12 in.) of sediment accumulation, the pond's bottom will need to be cleaned once every 18 years. This assumes that the bed load fraction is part of the reported TSS concentration and that there are no other sources of sediment, such as construction activities, being delivered to the ponds. Chances are that the actual accumulation rate will be somewhat higher and that more frequent cleanout will be needed.

Accumulation rates of heavy metals such as lead, zinc, copper, or other constituents may be a concern if such accumulations can create hazardous waste. If this is a concern, more frequent removal of sediments and periodic monitoring can be done to avoid these situations. Also, occasional core samples of pond or basin bottom will reveal if buildup of pollutants is occurring. If bottom sediment concentrations approach levels that would restrict disposal on site or in local landfills, site rehabilitation and total cleanout may be required.

Under existing U.S. EPA regulations (40 CFR Part 261), material cleaned from a detention pond should periodically be screened using the toxic characteristics leaching procedure (TCLP). This test should be carried out on accumulated sediment within the pond. If the sediment fails the test, it is subject to Resource Conservation and Recovery Act (RCRA) regulations and must be disposed of in an approved manner at a RCRA-approved facility. If the TCLP test is negative, sediments are subject to state and local solid waste disposal regulations.

If the material has been sufficiently dried to be a "workable material" and can pass a TCLP test, it can also be disposed of off site. This can be done at a landfill or as unclassified fill. However, sediments from any treatment facility can be nutrient-rich soils and, if other characteristics do not disqualify it, can be used in landscaping or as unclassified fill material. Disposing of accumulated sediment as fill or in landscaping avoids depleting landfill volume.

OTHER MAINTENANCE REQUIREMENTS. Mowing. Side slopes, embankments, emergency spillways, and other grassed areas of stormwater controls must be periodically mowed to prohibit woody growth and control weeds. More frequent mowing may be required in residential areas. Mowing can constitute the largest routine maintenance expense.

Debris and Litter Removal and Control. Debris and litter accumulate mostly near the inlet and outlet structures of stormwater controls and need to be removed during regular mowing operations. Particular attention should be paid to floatable debris that can eventually clog the outlet control structure or riser. Trash screens or trash racks can be strategically placed near inflow or outflow points to capture debris and assist with maintenance.

Litter and debris from illegal dumping should also be cleaned up on a regular basis, and an accurate log should be maintained of materials removed and improvements made. Controlling illegal dumping is difficult, but the posting of "no littering" or "no dumping" signs, with a phone number for reporting a violation in progress, may help. Adoption and enforcement of substantial penalties for illegal dumping and disposal could also be a deterrent.

Nuisance Control. Standing water or soggy conditions within a stormwater treatment control facility can create nuisance conditions for nearby residents. Odors, mosquitoes, weeds, and litter can all be potential problems in stormwater controls. However, wetland plants within a littoral zone of a retention pond (wet) provide a habitat for birds, predacious insects, and fish that serve as a natural check on mosquitoes. Also, regular maintenance to remove debris and ensure control structure functionality may also help control potential nuisance problems.

VEGETATED SWALES AND FILTER STRIPS

Biofilters consist primarily of vegetated swales and filter strips. Swales are shallow channels with flow depths often below the height of the vegetation that grows within them. Filter strips are vegetated flat surfaces over which water flows in a thin sheet. Planted vegetation can be turf grasses, emergent wetland, or high marsh plants. Some infiltration occurs through the underlying soil cover, but that is not the primary mode of treatment. Suspended solids are removed by filtering through the vegetation and settling. Dissolved constituents may also be removed through chemical or biological mechanisms mediated by the vegetation and the soil.

PLANNING CRITERIA AND GUIDELINES. Local governments and stormwater professionals should view biofilters as an element of the stormwater management infrastructure and as a part of the treatment train. For example, roadside ditches can be designed as biofilters and as landscaping amenities. Also, when land is limited, surrounding a pond with a biofilter will treat low flows before they enter the pond. Consider retrofitting biofilters in developed areas.

Effective biofiltration depends on proper design, construction, and maintenance. Also, local jurisdictions need to provide for access easements on private land for their inspection, monitoring, and maintenance, and they need to enforce long-term maintenance commitments by private parties to the control facilities owned by private parties.

Every effort should be made to prevent sediment-laden construction runoff, oil, and grease from entering biofilters. Catch basins, detention basins, presettling devices, and oil–water separators can be installed upstream of biofilters to help remove these materials before they reach the biofilter. Any of these devices or controls can improve the performance and the life of a biofilter.

DESIGN AND INSTALLATION CRITERIA AND GUIDELINES. Provisions Applying Equally to Swales and Filter Strips. It is important to maximize water contact with the biofilter vegetation and the soil surface. Graveled and coarse sandy soils are not desirable because they have difficulty sustaining vegetation. Heavy clay soils, materials toxic to vegetation, stones, and debris should also be avoided. When suitable, use on-site materials and scarify and till compacted soils before planting.

Vegetate biofilters uniformly with fine, turf-forming, water-resistant grasses. In arid and semiarid areas, supplemental irrigation will be needed to maintain healthy filter strips. Where biofilters intercept groundwater or where there is little slope for proper drainage, emergent herbaceous wetland vegetation is an acceptable planting alternative. Whenever possible, use vegetation native to the region. It is important to select grass and wetland species that work best for the region, climate, and native soils. Do not assume that information from other regions can be used at every site. Also, if wildlife habitat is being provided, select vegetation accordingly.

Use grass seed and mulch application rates specified by the supplier. If at all possible, do not use animal manure as an amendment and avoid using fertilizers. If fertilizer must be used, apply only the amount needed by the selected plants in the existing soil conditions and use a slow-release fertilizer. Establish grasses when natural moisture is adequate but irrigate if necessary. If wetland plants are used, they may need to be protected from predation with netting during establishment. If possible, divert runoff, other than necessary irrigation, during the period of vegetation establishment.

Vegetate upslope areas to prevent erosion. Use barrier shrubs to reduce intrusion by people and domestic animals. Avoid trees that shade biofilter grasses, and if trees cannot be avoided, space them at least 6.5 m (20 ft) apart. Landscape beds near biofilters should be at a slightly lower elevation than the adjacent ground surface.

Provide for a 1.0 to 2.0% slope in the direction of flow, with 6.0% being the maximum and 0.5% being the minimum. When the longitudinal slope is less than 1.0 to 2.0%, install a perforated underdrain or, if moisture is adequate, establish wetland species. If the slope is greater than 2.0%, use check dams to reduce the effective slope to approximately 2.0%. Install energy dissipating riprap at the toe and for a short distance downstream at the toe of these check dams to control erosion. When the land slopes more than 6%, swales can be installed to traverse the grade at a lesser slope. Grade biofilters carefully to attain uniform longitudinal and lateral slopes and eliminate high and low spots.

If curb cuts are used to distribute the flow over a biofilter, they should be at least 0.3 m (12 in.) wide to reduce clogging. Place the pavement slightly above the adjacent biofilter elevation. Install a flow-spreading device with sediment cleanouts (such as weirs, stilling basins, and perforated pipes) to uniformly distribute flow at an inlet to a swale or across the width of a filter strip. Protect inlet areas from erosion by using stilling basins and riprap pads with rock sized large enough not to be moved by the inflow.

A high-flow bypass is not needed if the biofilter is preceded by a runoff quantity control device designed to release flow at rates that will not cause erosion or scour within the biofilter. When a bypass is used, provide it with an inflow regulating device and use a pipe or a stabilized channel to convey the flow.

Provisions for Swales. At minimum, design for the peak runoff rate during the "maximized" storm. Base this storm depth on the rainfall depth using a runoff coefficient $C = 1.0$ (that is, complete runoff, no infiltration) and a 12-hour "drain time." Relate this "maximized" depth to an approximate intensity-duration curve for the area that can then be used with the rational formula by assuming the storm occurs over 1 or 2 hours. Using the time of concentration for the catchment and its runoff coefficient C, find the design flow rate. Unless larger events will bypass the swale, enlarge its capacity for flood passage of the 10- to 100-year peak flow.

The following criteria are probably most applicable in warm and temperate, non-semi-arid climates and should be met or exceeded during the biofiltration capacity design event:

- "Maximized" runoff hydraulic residence time of 5 minutes or more;
- Maximum flow velocity less than 0.3 m/s (0.9 ft/sec);
- Manning's $n = 0.20$ for routinely mowed swales;
- Manning's $n = 0.24$ for infrequently mowed swales;
- Maximum bottom width of 2.4 m (8 ft);
- Minimum bottom width of approximately 0.6 m (2 ft);
- Maximum depth of flow no greater than one-third of the gross or emergent wetland vegetation height for infrequently moved swales or greater than one-half of the vegetation height for regularly mowed swales, up to a maximum of approximately 75 mm (3 in.) for grass and approximately 50 mm (2 in.) below the normal height of the shortest wetland plant species in the biofilter; and
- Minimum length of 30 m (100 ft).

Use a trapezoidal cross section for ease of construction with side slopes no steeper than 4:1 for ease of maintenance. Terracing needs to be used when side slopes become steeper than 3:1.

Provisions for Filter Strips. Design filter strips to carry the "maximized" storm peak runoff rate described in the section titled Provisions for Swales. Base the flow capacity design to meet or exceed the following criteria during the design storm event. The following specifications may be used:

- Hydraulic retention time of no fewer than 5 minutes;
- Average velocity not greater than 0.3 m/s (0.9 ft/sec);
- Manning's $n = 0.20$ for routinely mowed strips;
- Manning's $n = 0.24$ for infrequently mowed strips;
- Limit filter strip width to achieve uniform flow distribution;
- Average depth of flow no more than 25 mm (1.0 in.); and
- Hydraulic radius taken to be equal to the design flow depth.

In arid and semiarid areas, biofilter strips will need irrigation to maintain a dependable grass cover.

OPERATION AND MAINTENANCE OF BIOFILTERS. To keep biofilters operating properly, keep all inlet flow spreaders even and free of debris. Remove debris for aesthetic reasons. Mow grass-covered biofilters regularly during the growing season to promote growth and pollutant uptake. Remove cuttings and dispose of them properly or through composting. If the objective is to prevent nutrient transport, mow grasses or cut emergent wetland plants to a low height, but still above the maximum flow depth, at the end of the growing season. For trapping floatables and debris pollution control objectives, let the plants stand at a height exceeding the design water depth by at least 50 mm (2 in.) at the end of the growing season.

Remove sediment by hand with a flat-bottomed shovel during the summer months whenever sediment covers vegetation or begins to reduce the biofilter's capacity. Reseed damaged or recently maintained areas immediately with a mix used for initial establishment or use grass plugs from adjacent upslope areas. If possible, redirect flow until new grass is firmly established. Otherwise, cover the seeded areas with a high-quality erosion control fabric.

Inspect biofilters periodically, preferably monthly, especially after heavy runoff. Maintain clean curb cuts to avoid soil and vegetation buildup. Educate local residents about the importance of keeping biofilters free of lawn debris and pet waste. Base roadside ditch cleaning on hydraulic analysis. Remove only the amount of sediment necessary to restore needed hydraulic capacity, leaving as much of the vegetation in place as possible. Eventually, sufficient sediment will be trapped that the entire biofilter will need to be removed with the sediment and reconstructed to begin a new cycle of stormwater quality control.

BIOFILTER DESIGN PROCEDURE. Preliminary Steps. The design procedures for swales and filter strips are almost identical. The design steps described below are for swales, with notes provided for filter strip design when appropriate:

- P1: estimate runoff flow rate for the design event and limit the discharge to approximately 0.03 m^3/s (1.0 cu ft/sec) by dividing the flow among several swales, installing upstream detention to control release rates, or reducing the developed surface area to reduce runoff coefficient and gain space for biofiltration.
- P2: establish the slope of the proposed biofilter.
- P3: select a vegetation cover suitable for the site.

Design for Biofiltration Capacity. This analysis emphasizes biofiltration rather than efficient hydraulic conveyance, thereby promoting sedimentation, filtration, and other pollutant removal mechanisms. Typically, a lower maximum velocity is arrived at than required for slope stability, and the biofilter dimensions typically do not have to be modified after a check for stability.

- D1: estimate the height of vegetation that is expected to occur during the storm runoff season. The design flow depth should be at least 50 mm (2.0 in.) less than this vegetation height and a maximum of approximately 75 mm (3.0 in.) in swales and 25 mm (1.0 in.) in filter strips.

- D2: select Manning's n as discussed.
- D3: typically swales are designed as trapezoidal channels (skip this step in filter strip design). When using a rectangular section, provide reinforced vertical walls. See Figure 5.5 for relationships for estimating cross-sectional area, top width, and hydraulic radius for typical channel geometrics.

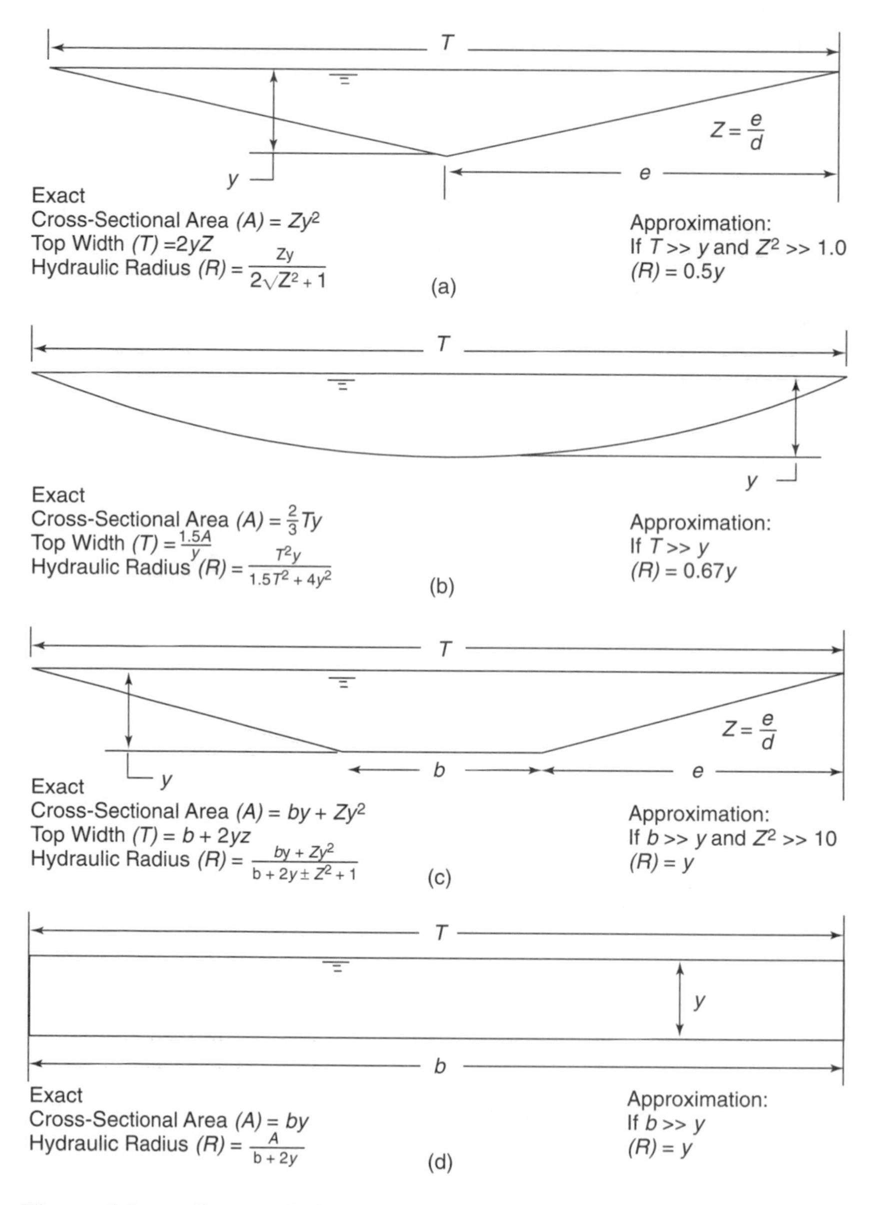

Figure 5.5 Geometric formulas for common swale shapes: (a) v-shape, (b) parabolic shape, (c) trapezoidal shape, and (d) rectangular shape.

- D4: use Manning's equation to approximate initial dimensions. For trapezoidal shape, select a side slope that is no steeper than 3:1, with 4:1 or flatter preferred. Set the bottom width to be between approximately 0.6 and 2.5 m (2.0 to 8.0 ft).
- D5: compute cross-sectional area.
- D6: compute the flow velocity for the design flow rate. Limit velocities during the "maximized" design storm to less than 0.3 m/s (0.9 ft/sec). Greater velocities were found to knock grasses from a vertical position in the Pacific Northwest (U.S.), reducing filtration. Experience in other regions may be different. Adjust for local experience. If the flow velocity exceeds the limit value, repeat steps P1 through D6.
- D7: compute the swale length using the design velocity from step D6 and an assumed hydraulic retention time. A suggested retention time value in the Pacific Northwest is 9 minutes. However, it is acceptable to use other accepted regional values, preferably no fewer than 5 minutes. If the computed swale length is less than 30 m (100 ft), increase to 30 m and adjust bottom width.

Check for Stability to Reduce Erosion. The "stability" check is performed for the combination of highest expected flow and least vegetation coverage and height.

- S1: unless runoff will bypass the biofilter, perform the "stability" check for the 10- to 100-year design storm. Estimate the design discharge as in step P1.
- S2: estimate the vegetation coverage (for example, "good" or "fair") and flow depth for conditions that will exist whenever the coverage and vegetation height are the least.
- S3: estimate the degree of retardance using Table 5.7. Because emergent wetland species typically grow less densely than grasses, assume a "fair" coverage.
- S4: establish the maximum permissible velocity (V_{max}) from Table 5.8 to prevent erosion.

Table 5.7 Guide for selecting degree of retardance.

Average grass height, in.[a]	Degree of vegetation coverage	
	"Good"	"Fair"
>30	A. (Very high)	B. (High)
11–24	B. (High)	C. (Moderate)
6–10	C. (Moderate)	D. (Low)
2–6	D. (Low)	D. (Low)
<2	E. (Very low)	E. (Very low)

[a] in. × 25.40 = mm.

Table 5.8 Recommended maximum velocities for swale stability.

Cover	Slope, %	Maximum velocity, ft/sec[a]	
		Erosion-resistant soils	**Easily eroded soils**
Kentucky bluegrass	0–5	6	
Tall fescue			5
Kentucky bluegrass	5–10	5	4
Ryegrass			
Western wheat-grass			
Grass–legume mix	0–5	5	4
	5–10	4	3
Red fescue	0–5	3	2.5
Redtop	5–10	NR[b]	NR[b]

[a] ft/sec × 0.304 8 = m/s.
[b] Not recommended.

- S5: select a trial Manning's n. A reasonable initial choice under poor vegetation cover conditions is $n = 0.04$.
- S6: use Figure 5.6 to help approximate the value for VR (that is the product of velocity and hydraulic radius) using retardation information from step S3 and V_{max} from step S4.
- S7: estimate a new hydraulic radius by dividing (VR) from step S6 by V_{max} from step S4.
- S8: use Manning's equation to solve for the actual VR.
- S9: compare the actual VR from step S8 against the first approximation in step S6. If they do not agree within 10%, repeat steps S5 through S9 until acceptable agreement is reached.
- S10: compute the actual V for the final design conditions and check to be sure that $V < V_{max}$, from step S4.
- S11: compute the cross-sectional area for "stability."
- S12: compare the area computed in step S11 with the area computed from the biofiltration "capacity" analysis (step D5). If less area is required for "stability" than is provided for "capacity," the "capacity" design is acceptable. If not, use area from step S11 and recalculate channel dimensions.
- S13: calculate the depth of flow at the "stability" check design flow rate condition for the final dimensions.
- S14: compare the depth from step S13 to the depth used in the biofiltration "capacity" design. Use the larger of the two and add 0.3 m (1.0 ft) freeboard. Calculate the top width for the full depth. Skip this step in filter strip design.
- S15: make a final check for discharge capacity based on the "stability" check design storm and maximum vegetation height and cover (this

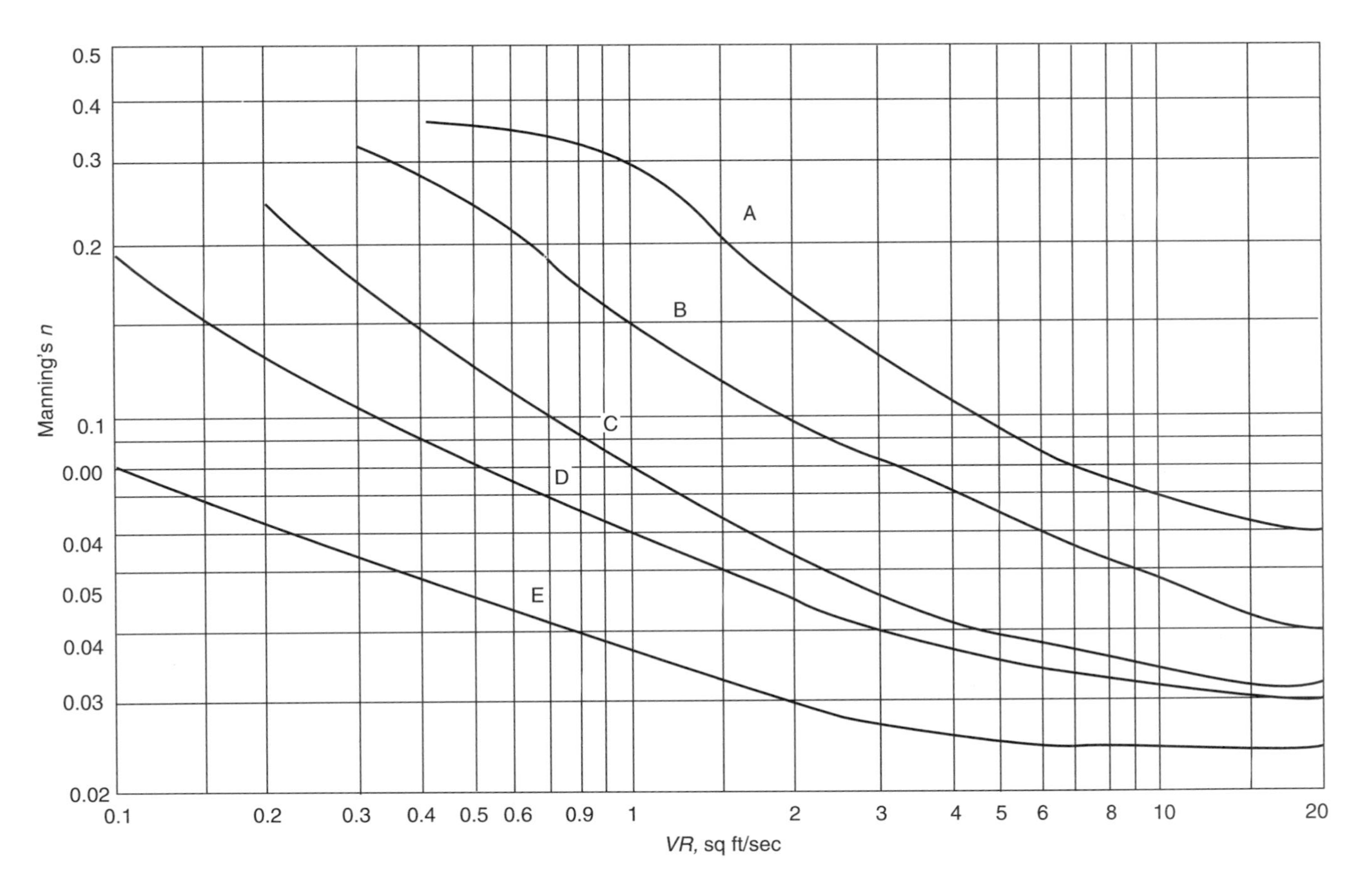

Figure 5.6 **Manning's *n* versus *VR* versus degrees of flow retardance (note that *VR* is the product of velocity and hydraulic radius) (sq ft/sec × 0.092 9 = m^2/s).**

check will ensure capacity is adequate if the largest expected event coincides with the greatest retardance). Using Manning's *n* selected in step D2 and the calculated channel dimensions (including freeboard), and compute the flow capacity of the channel under these conditions. If this flow capacity is less than the "stability," check design storm flow rate, increase the channel cross-sectional area as appropriate, and repeat calculations. Specify the new channel dimensions.

Completion Step. Review guidelines provided in Sections 1, 2, and 3 for biofilter planning, design, installation, and operation, and specify all appropriate features applicable for installation.

STORMWATER INFILTRATION AND PERCOLATION

Whenever site conditions permit, a portion of urban stormwater runoff can be disposed of locally through infiltration. Surface infiltration can be encouraged to occur through the use of grass buffer strips, swales, porous pavement, and dedicated infiltration basins. Stormwater runoff can also infiltrate to the ground near its origin through percolation trenches similar to those used in wastewater septic system leach fields.

Although surface infiltration has received much attention in recent years, it has long been in use and was practiced because good site drainage was lacking. The use of "inefficient" surface runoff in urban areas through the use of permeable vegetated surfaces has gained considerable but not universal acceptance in recent years.

The practice of minimizing directly connected impervious area (MDCIA) is a practice that is most appropriate for sites that are planning to have less than 60% imperviousness with relatively flat terrain. When properly designed and used, effective site drainage can still be provided. Figure 5.7 shows a comparison between traditional and MDCIA land development practices. Note that this is done by reducing the amount of directly connected impervious area, directing all roof downspouts onto lawns instead of to the curb and gutter, using stabilized road shoulders and swales, using modular block porous pavement whenever feasible, and maximizing the lengths of the surface paths leading to formal infiltration areas.

Formal on-site stormwater infiltration is accomplished through the use of infiltration basins and percolation trenches. These facilities need to be designed to also provide good surface and subsurface site drainage. On-site disposal is most successful when used to control runoff from individual building sites and small urban catchments (that is, up to 4 ha [10 ac] of single family residential and up to 2 ha [5 ac] of commercial lands).

Infiltration and percolation facilities are susceptible to early failure. Fail-

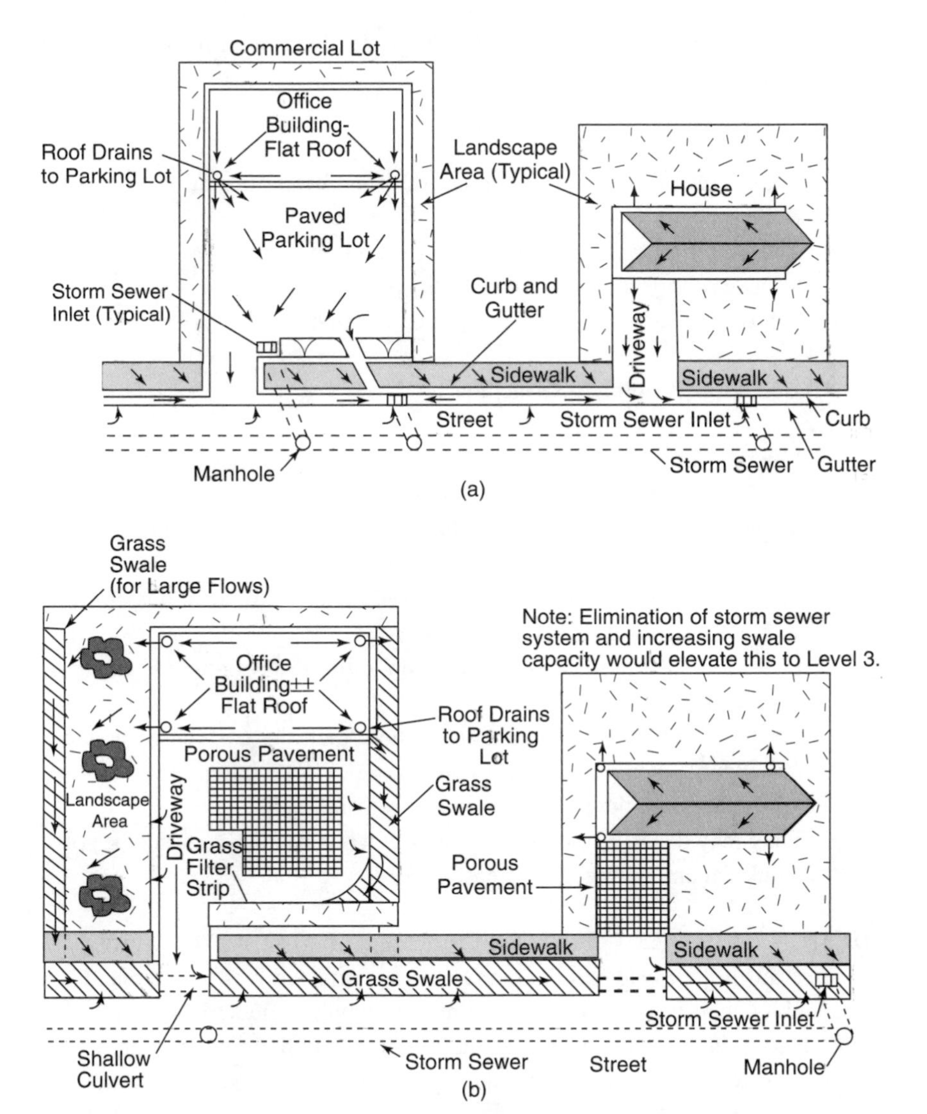

Figure 5.7 **Land development practice: (a) traditional site and street drainage design and (b) limiting directly connected impervious areas (UDFCD, 1992).**

ure is defined when stormwater no longer drains into the ground at design rates. Early signs of failure include excessively long periods of standing water on infiltrating surfaces or within percolation trenches. With failure, downstream stormwater systems and receiving waters experience increased higher runoff rates and greater volumes on a more frequent basis with corresponding pollutant loads.

Infiltration basins and percolation trenches can only be used at sites with porous soils, favorable site geology, and proper groundwater conditions. They should not be used when local institutions cannot ensure their proper installation and long-term maintenance. These infiltration basins and percolation trenches may

- Recharge groundwater,
- Reduce ground settlement in areas of groundwater depletion,
- Help to preserve and enhance local vegetation,
- Reduce pollutant loads transported to receiving waters,
- Reduce runoff volumes and peak flows, and
- Allow the use of smaller storm sewer systems.

Arguments against their use include

- Much of the surface runoff occurs from publicly owned streets and large commercial areas—infiltration and percolation facilities located on individually owned lots may have little effect on runoff rates and volumes experienced downstream;
- Infiltrating surfaces and soils seal as sediment and other solids clog soil pores, causing their failure;
- Individual infiltration and percolation facilities may not receive proper maintenance;
- Local groundwater mounding under these facilities can cause them to fail and may damage adjacent structures and flood basements;
- When they fail, they are expensive to reinstall; and
- They may degrade groundwater quality when used in certain types of industrial and commercial areas.

DESIGN CAPTURE VOLUME. The amount of urban surface runoff arriving at infiltration and percolation facilities is affected by the watershed's size and its imperviousness, local rainfall, snowmelt characteristics, and other factors such as lawn watering, car washing, and hydrant flushing. Of these, stormwater and snowmelt runoff typically are the most significant considerations for design, and the choice of a design is often dictated by local conditions or criteria. It is recommended that infiltration basins and percolation trenches be sized to handle no less than the "maximized" storm runoff volume described earlier in this chapter. Base the design volume for infiltration basins on the coefficients listed for the event capture ratio, 12-hour drain time, and the tributary catchment's runoff coefficient. Estimating the design volume for percolation trenches is somewhat more complex and is described later in this chapter.

SNOWMELT. In some locations, snowmelt can govern the size of infiltration and percolation devices, especially when the tributary catchment has little impervious area and much pervious area. Under extreme conditions, snowmelt rates can equal 4 mm (0.6 in.) of water per hour. Although it is not possible to typify snowmelt runoff rates from across the U.S., the following snowmelt rates may be used to check if snowmelt governs the size of these facilities: impervious surfaces—1.0 mm/h (0.04 in./hr), and pervious surfaces—0.5 mm/h (0.02 in./hr). Use local rates whenever available.

SURFACE INFILTRATION BASINS. Screening for Site Suitability. There are several conditions that can be used to eliminate a site for local stormwater infiltration. These include the following:

- Seasonal high groundwater is less than 1.2 m (4 ft) below the infiltrating surface;
- Bedrock or an impervious soil layer is within 1.2 m (4 ft) of the infiltrating surface;
- The infiltrating surfaces are located on top of fill or compacted soils; or
- The surface and underlying soils are Resource Conservation Service (RCS) hydrologic soil group C or D, or the saturated surface infiltration rate is less than 7.6 mm/h (0.3 in./hr).

If the above conditions do not rule out the new development as a candidate for infiltration, assess its suitability using a point evaluation system suggested by the Swedish Association of Water and Wastewater Works (1983). This assessment technique, when used properly and objectively, has been proven to be effective. It was first described in the U.S. by Urbonas and Stahre (1993) and is based on assessing various site conditions by assigning points for each category listed in Table 5.9.

A site with fewer than 20 points is considered unsuitable, while a site with more than 30 points is considered good. A site with 20 to 30 points is considered to be a fair candidate, with occasional standing water on the infiltration surfaces likely. This preliminary screening technique, however, is not a substitute for detailed site-specific engineering investigations. When the initial screening process finds the site acceptable, the infiltration surface area and the stormwater storage volume above this surface must then be determined. Note that Table 5.9 suggests that the surface area of all infiltration areas within a development (including buffer strips, lawns, and swales) be no less than one-half of the tributary impervious surface areas.

This screening procedure can best be illustrated by an example. For example, an infiltration site is to dispose of stormwater runoff from a roof having a 1.0-ha (2.5-ac) area. The site is a new lawn with a surface area of 1.0 ha (2.5 ac) and a 0.20% slope. The topsoil is normal humus, and the underlying soils are composed mostly of coarse silt. Determine if the site is suitable for infiltration.

Using the evaluation point system presented in Table 5.9, the results are as follows:

- The infiltrating area is 1.6 times larger than the impervious surface (that is, $A_{INF} = 1.6\,A_{IMP}$) = 10 points.

Table 5.9 **Point system for the evaluation of potential infiltration sites (STORMWATER—BEST MANAGEMENT PRACTICES by URBONAS/STAHRE, © 1993. Adapted by permission of Prentice-Hall, Inc., Upper Saddle River, N.J.).**

Site condition	Evaluation points to award
Ratio of tributary impervious area (A_{IMP}) to the infiltrating surface area (A_{INF})	
$A_{INF} > 2 \cdot A_{IMP}$ 20 points	10 points
$A_{IMP} < A_{INF} < 2 \cdot A_{IMP}$	5 points
$0.5 \cdot A_{IMP} < A_{INF} < A_{IMP}$	0 points
Urban catchments with pervious surface areas less than one-half of the impervious surfaces are poor candidates	
Surface soil layer type	
Coarse soils with low organic material content	7 points
Normal humus soil	5 points
Fine-grained soils with high ratio of organic matter	0 points
Underlying soils	
If the underlying soils are more coarse than surface soil, assign the same number of points for underlying soils as were given for the surface soil layer soils above	
If the underlying soils are finer grained than the surface soils, use the following points:	
Gravel, sand of glacial till with gravel or sand	7 points
Silty sand or loam	5 points
Fine silt or clay	0 points
Slope (S) of the infiltration basin's site	
$S < 0.007$ ft/ft (m/m)	5 points
$0.007 < S < 0.020$ ft/ft (m/m)	3 points
$S > 0.020$ m/m	0 points
Vegetation cover	
Healthy natural vegetation cover	5 points
Lawn well established	3 points
Lawn new	0 points
No vegetation, bare ground	−5 points
Degree of traffic on infiltration surface	
Limited foot traffic	5 points
Average foot traffic (park, lawn)	3 points
Much foot traffic (playing fields)	0 points

- The top soil layer is of normal humus type = 5 points.
- The underlying soil layers are coarse silt = 5 points.
- The slope of the infiltration surface is 0.002 = 5 points.
- The infiltration surface is a new established lawn = 0 points.
- The lawn is expected to have normal foot traffic = 3 points.
- The total is 28 points for this site. This site is judged to be an above-average candidate, runoff is not likely to puddle frequently, and occasional periods of standing water are likely.

Configuring a Basin. Infiltration basins need to empty through the basin's bottom within a reasonably short period. Otherwise, boggy and undesirable site conditions will occur, and the grasses lining these basins will die. Size the basin's surface area to drain the design capture volume in fewer than 6 hours. Also, design the basin with a shallow maximum ponding depth. An ideal site will not result in water ponding more than 0.3 m (1 ft) deep during the design storm, with the excess volume either overflowing or bypassing the basin. Recognize that the available soil pore volume will amplify each 0.3 m (1 ft) of ponding depth into 0.9 to 1.2 m (3 to 4 ft) of groundwater depth under the basin, which water column then needs to drain off laterally.

There is a strong possibility that unless the underlying soils have high hydraulic conductivity, it will take few runoff events to create a groundwater mound under a basin. In many soils, groundwater mounding drains off laterally slowly and can surface, causing a failure. These failures can be reduced if the development's infiltration is distributed uniformly throughout the development site. Using many infiltration basins distributed throughout the development tends to more closely reproduce the predevelopment condition. Thus, instead of concentrating all site runoff at one infiltration basin, it is better to install many small infiltration basins throughout the development site. Try to fit them into the landscape, even into individual residential or commercial lots.

Vegetate all infiltrating surfaces with grasses that can withstand and survive prolonged periods of inundation, followed by extended dry periods. Healthy vegetation is essential—without it, the surface soil pores quickly seal. Grass roots are needed to reopen them and will help to do so even when considerable deposition of silt has occurred. Eventually, the deposited sediment layer and old grass will need to be removed, the soils rehabilitated, and the basin revegetated.

Sizing Infiltration Basins. An infiltration basin cannot transfer stormwater to the ground as rapidly as stormwater arrives at the basin. As a result, a detention volume is needed above all infiltration surfaces to temporarily store this excess runoff. Table 5.10 contains a list of the infiltration rates for several typical soil groups. It is best, however, to perform several surface infiltration tests at each site and use an average of several of the lowest measured infiltration rates for design purposes. It is also important to recognize that as

Table 5.10 Typical infiltration rates of various soil groups.

Soil conservation service group	Typical soil type	Saturated infiltration rate	
		mm/h	in./hr
A	Sand	200	8.0
A	Loamy sand	50	2.0
B	Sandy loam	25	1.0
B	Loam	12.7[a]	0.5[a]
C	Silt loam	6.3[a]	0.25[a]
C	Sandy clay loam	3.8	0.15
D	Clay loam and silty clay loam	<2.3	<0.09
D	Clay	<1.3	<0.05

[a] Minimum acceptable infiltration rate is 7.6 mm/h (0.3 in./hr). Sites with soils with lesser rates should not be used.

sediment accumulates on the basin's bottom, the effective infiltration rate will be governed by the sediment layer, which in turn will be affected by the presence or absence of a healthy grass surface. For this reason, it is suggested that when local soils exhibit high surface infiltration rates, the basin's design be based on infiltration rates that do not exceed 50 mm/h (2.0 in./hr). At the same time, when native soils have infiltration rates less than 50 mm/h (2 in./hr), the designer should consider using a somewhat reduced rate to account for the fact that soil infiltration rates will decline as sediment builds up on the bottom of the basin.

The detention volume of an infiltration basin is found using Equation 5.2. Use the "maximized" stormwater capture volume based on a 12-hour drain time. Next, estimate the basin's surface area using a maximum detention depth of 0.3 m (1 ft). The known soil's infiltration rate is then multiplied by the basin's surface area to find the exfiltration rate. Ascertain that the detention volume will drain in 6 hours or fewer. If it does not, increase the surface area until the volume drains in fewer than 6 hours. Last, check if the basin area has to be increased to handle snowmelt.

For example, use a 2.22-ha (5.5-ac) catchment in Minneapolis, Minnesota, located on sandy loam soils with saturated infiltration rate of 25 mm/h (1.0 in./hr). The catchment is 44% impervious (that is, runoff coefficient $C = 0.3$).

Using Equation 5.2 and Table 5.4, the "maximized" volume for a detention basin with a 12-hour drain time is

$$P_0 = 1.109 \times 0.5 \times 0.3 = 4.2 \text{ mm (0.166 watershed in.)}$$

The design volume then is

$$V_{WQ} = (0.166 \div 12) \times 5.5 = 0.076 \text{ ac-ft} = 94 \text{ m}^3 \text{ (3 320 cu ft)}$$

Limiting the ponding depth to 0.3 m (1 ft) establishes the basin's surface area at 308 m^3 (3 320 sq ft). The total exfiltration rate then is

$$Q_{out} = (3\,320 \times 1.0 \div 12) = 7.8\ m^3/h\ (277\ cu\ ft/hr)$$

This exfiltration rate will empty the design volume in 12 hours. Doubling the basin's surface area to 716 m^2 (6 640 sq ft) will empty the design volume in 6 hours at a rate of 15.7 m^3/h (554 cu ft/hr). Note that the resultant basin area occupies almost 3% of the total catchment area.

Next, check to see if the basin will handle prolonged snowmelt periods without overtopping. Using the snowmelt rates listed previously, we find the snowmelt rate for the site is 0.71 mm/h (0.028 in./hr), equal to a 15.8-m^3/h (55-cu ft/hr) runoff rate. This is virtually identical to the design rate of 15.7 m^3/h (554 cu ft/hr), and further adjustment to the basin's size is not justified.

PERCOLATION TRENCHES. Assessing a Site for Suitability. Darcy's law provides a basis for estimating the rate at which water can percolate into the ground through the sides of a percolation trench. It is expressed as Equation 5.4 and forms a basis for judging whether a site is suitable for the installation of a percolation trench:

$$U = k \cdot I \tag{5.4}$$

Where

U = flow velocity, m/s;
k = hydraulic conductivity, m/s; and
I = hydraulic gradient, m/m.

Because the bottom of the facility is above the high seasonal groundwater, assume the hydraulic gradient to be $I = 1.0$. Determine the hydraulic conductivity of the soils adjacent to the percolation trench. Table 5.11 lists ranges in hydraulic conductivity for a variety of typically found soil types. Note that conductivities can vary four orders of magnitude for a single soil group. It is best to perform several site-specific hydraulic conductivity tests and use the

Table 5.11 Hydraulic conductivity of five soil types.

Soil type	Hydraulic conductivity	
	ft/sec	m/s
Gravel	3.3×10^{-3}–3.3×10^{-1}	10^{-3}–10^{-1}
Sand[a]	3.3×10^{-5}–3.3×10^{-2}	10^{-5}–10^{-2}
Silt	3.3×10^{-9}–3.3×10^{-5}	10^{-9}–10^{-5}
Clay (saturated)	$<3.3 \times 10^{-9}$	$<10^{-9}$
Till	3.3×10^{-10}–3.3×10^{-6}	10^{-10}–10^{-6}

[a] Minimum acceptable hydraulic conductivity for stormwater percolation is 2.0×10^{-5} m/s (6.5×10^{-5} ft/sec).

lowest field measured *in situ* hydraulic conductivities for final design purposes. Even under ideal conditions, soils adjacent to the trench will clog with time. A percolation trench is expensive to construct and more expensive to rebuild. Thus, being conservative in its design is appropriate.

The same factors, except for the use of soil hydraulic conductivity, affect site suitability for a percolation facility as affect a surface infiltration basin. Thus, if the following conditions are discovered or are likely to be at the site, disposal of stormwater by percolation is not recommended:

- Seasonal high groundwater is less than 1.2 m (4 ft) below the bottom of the percolation trench;
- Bedrock or impervious soils are within 1.2 m (4 ft) of the bottom of the percolation trench;
- The percolation trench is located within or on top of fill or recompacted soils; or
- The soils adjacent to the trench are RCS hydrologic group C or D or the field saturated hydraulic conductivity of the soils is less than 2.0×10^{-5} m/s (6.5×10^{-5} ft/sec).

If the above conditions do not rule out the site, the Swedish Association of Water and Wastewater Works (1983) provides design recommendations. This procedure is described by Urbonas and Stahre (1993).

Configuring a Percolation Trench. Percolation trench design uses the pore volume of trench fill media as the detention volume. Table 5.12 lists the porosity of the more typical trench fill materials. The bottoms of these trenches tend to clog first, often shortly after installation. As a result, the bottom of the trench is considered impervious and all water is assumed to percolate out only through its walls. Typically, long and deep trenches are most efficient and require the least amount of porous media. The maximum trench depth is limited by trench-wall stability, seasonal high groundwater levels, and the depth to any impervious soil layer. Trenches 1 m (3 ft) wide and 1 to 2 m (3 to 6 ft) deep seem to be most efficient.

If a percolation trench cannot be made sufficiently large to empty its fully available water storage volume (that is, granular media pore space volume)

Table 5.12 Porosity of commonly used granular materials.

Material	Effective porosity, %
Crushed and blasted rock	30
Uniform sized gravel	40
Graded gravel, 2.0 cm (0.75 in.)	30
Sand	25
Pit run gravel	15–25

within a 24-hour period, it is recommended that a collector pipe be installed near its bottom and the stored water be released slowly through a flow controller (that is, choked outlet). The outlet is designed to supplement the percolation outflow so that both combine to empty out the trench-full volume in 24 hours. This type of installation behaves, in part, like a detention basin.

Most important for percolation trench longevity is to filter all the stormwater entering the trench through a sand layer. Percolation trenches should not be used without first filtering its inflow or if these filter systems will not be adequately maintained. If stormwater is permitted to enter the trench without first being filtered, pore media and adjacent soils will seal with time and the facility will fail.

Figure 5.8 illustrates a percolation trench with a surface sand filter layer

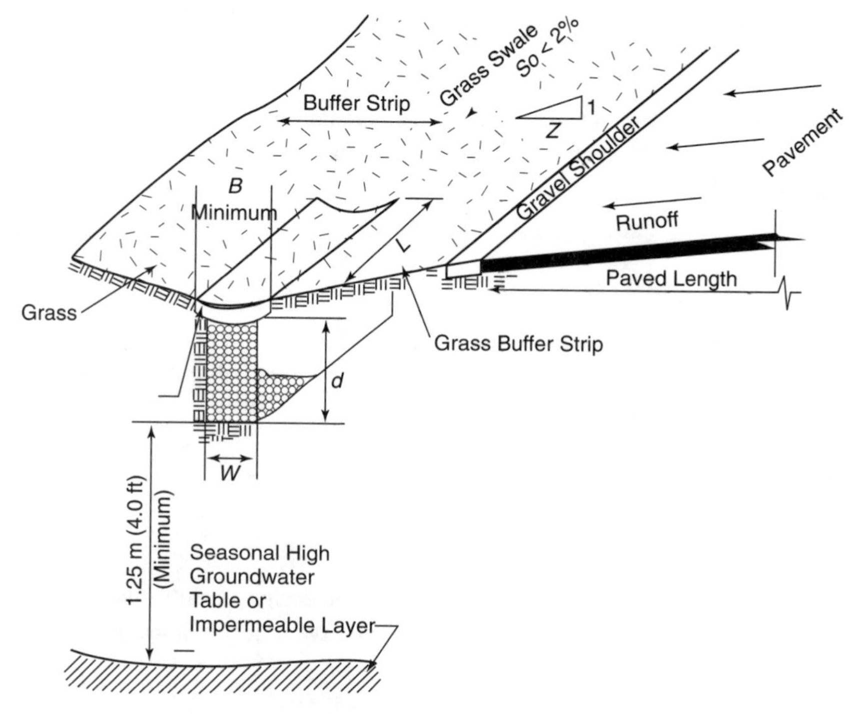

Figure 5.8 A percolation trench with a sand filter layer for surface inflow (STORMWATER: BEST MANAGEMENT PRACTICES BY URBONAS/STAHRE, © 1993. Adapted by permission of Prentice-Hall, Inc., Upper Saddle River, N.J.). (Notes: Wrap all rock fill in geotextile fabric with coarse pores; buffer strip length > 25% paved length; length $L \leq K/(So/100)$ in which $K = 0.3$ m (1 ft) and So = slope, %; add trenches as needed to obtain required total length for infiltration; side slope $Z \geq 4.0$; sand or sand–turf filter layer surface area shall be sized to permit inflow to the trench with minimum of ponding; create ponding on sand–turf layer using a berm across the swale; $B > 2d$.)

on top. This sand filter layer has to have sufficient surface area to permit stormwater to enter the trench with minimal ponding above it. However, some ponding volume will be needed above the sand filter to buffer higher rates of runoff. Such a sand surface filter layer can be a modular porous pavement. Other filter configurations are possible, including inlets with geotextile filter bags within them (Urbonas and Stahre, 1993). All filter devices will need aggressive routine maintenance for acceptable operation.

Unlike the surface infiltration facilities, failure of a percolation trench can be unnoticed for long periods of time because the trench is out of sight. A routine inspection program is needed to discover failed percolation trenches. It is unlikely that someone will randomly observe and report a trench failure during storm periods. One or more observation wells should be provided to facilitate inspections. A record of water in the trench not draining within 2 days after a wet period ends can indicate incipient failure and should be investigated.

Sizing a Percolation Trench. Because a percolation trench is used to limit runoff from a small catchment, rational formula (Equation 5.5) may be used for its design:

$$Q = K_u \cdot C \cdot I_t \cdot A \tag{5.5}$$

Where

Q = average runoff rate for the storm duration t, m^3/s (cu ft/sec);
K_u = unit conversion factor, 1.0 for U.S. standard units (36 for the International System of Units);
C = runoff coefficient, nondimensional;
I_t = rainfall intensity for the design storm at the storm duration t, mm/h (in./hr); and
A = area of the tributary watershed, ha (ac).

Multiplying the average runoff rate, Q, by the design storm's duration, t, results in Equation 5.6, which gives the cumulative runoff volume over time t:

$$V_{in}(t) = K \cdot 3\,600 \cdot C \cdot \frac{I_t}{1\,000} \cdot t \cdot A \tag{5.6}$$

Where

$V_{in}(t)$ = total volume of *inflow* over storm duration t, m^3 (cu ft); and
t = storm duration, hour.

Because the water depth in the trench varies during storm runoff, the sides of a percolation trench are not fully inundated during the runoff event. To simplify the sizing process, the designer can assume that the average outflow rate is the result of one-half of the trench depth being inundated. This then allows the designer to find the average effective area of percolation. Also, assume the hydraulic gradient, I, equals 1.0. Thus, Equation 5.7 is derived from

Darcy's law (that is, Equation 5.4):

$$V_{out}(t) = 3\ 600 \cdot k \cdot (A_{perc} \div 2) \cdot t \quad (5.7)$$

Where

$V_{out}(t)$ = total volume of water percolated into the ground over time t, m^3;
k = hydraulic conductivity of the soil, m/s (ft/sec);
A_{perc} = total area of the sides of the percolation trench, m^2 (sq ft);
t = duration of the percolation time, hour.

The maximum volume of water stored, V, in the trench is the difference between $V_{in}(t)$ and $V_{out}(t)$, as expressed by Equation 5.8:

$$V = \max\ [V_{in}\ (t) - V_{out}\ (t)] \quad (5.8)$$

Thus, Equation 5.9 is derived by combining Equations 5.5 and 5.7 into Equation 5.8:

$$V = \max\ [3\ 600 \cdot K_u \cdot I_t \cdot C \cdot A \cdot t - 1\ 800 \cdot k \cdot A_{perc} \cdot t] \quad (5.9)$$

Configure the trench to drain the "maximized" storm volume discussed earlier in the chapter. First, find this volume using Equation 5.2 for the 12-hour drain time and a runoff coefficient $C = 1.0$. The trench is designed to dispose of the runoff from such a storm through percolation through the sides. Because the detention time in the trench is not the issue for water quality enhancement, the maximized depth is used to define the intensity-duration function of a design storm by assuming this rainfall depth occurs within 1 hour.

The next step is to select a cross section for the trench and the type of fill material. Assume a trench length and test it for adequacy. Eventually, through a trial-and-error process, the assumed trench length agrees with the calculated one. This procedure can be reduced to a spreadsheet to facilitate the iterative solution. Figure 5.9 presents an example of such a spreadsheet. After the known parameters are entered, the iterative process begins by entering an assumed trench length and calculating the "needed trench length." New "assumed length" values are entered until a balance is achieved between the "assumed" and "needed" lengths. In this example (that is, Figure 5.8) the needed trench length was found to be 44 m (144 ft).

OTHER INFILTRATION FACILITIES. Buffer Strips and Swales. The design for buffer strips and swales is described earlier in this chapter. Both buffer strips and swales can infiltrate stormwater to the ground. However, the duration of time that runoff actually is in contact with these surfaces is relatively short, and, as a result, the volume of infiltration is limited. Nevertheless, buffer strips and swales can infiltrate significant fractions of the smaller runoff events when they are located on porous soils. Their use is encouraged,

Project Title: Percolation Trench Sizing

Tributary Catchment Area [A]:	**5.50** ac
Percent Impervious:	**44.0%**
Runoff Coefficient [$C = 0.858\ i\hat{}3 - 0.78\ i\hat{}2 + 0.774\ i + 0.04$]:	0.30
Maximized Rainfall Depth ($I_{1\text{-hour}}$); *$C = 1.0$ and 12-hour down time:*	**0.50** in.
Soil's Hydraulic Conductivity:	**0.001** ft/sec
Trench Width (W):	**3.00** ft
Height (H):	**6.00** ft
Assumed Length (L):	***144.0*** ft (trial length)*
Hydraulic Gradient:	**1.00**
Average Percolation Outflow Rate {$Q_{out} = kH(L + W)$}:	0.882 cu ft/sec
Rock Media Porosity (p):	**0.35**

Note: All values in bold typeface are user input values.

Rainfall Intensity I-D Curve Value at Duration T: $I = \{a{*}I_{1\text{-hour}}/(T + b)\hat{}c\}$

Local Coefficient a = **28.5** b = **10.00**

c = **0.786**

Storm duration, min.	Rainfall intensity, in./hr	Runoff volume, cu ft	Outflow volume, cu ft	Volume stored, cu ft	Needed trench volume, cu ft	Needed trench length, ft
T (1)	I (2)	*60 CIAT* (3)	*60 Q_{out} T* (4)	*(3)–(4)* (5)	*(5)/p* (6)	*(6)/(WH)* (7)
0.0	2.33	0	0	0	0	0
10.0	1.35	1 351	529	822	2 348	130
20.0	0.98	1 965	1 058	906	2 589	144*
30.0	0.78	2 350	1 588	763	2 180	121
40.0	0.66	2 630	2 117	513	1 466	81
50.0	0.57	2 848	2 646	202	578	32
60.0	0.51	3 028	3 175	−147	−421	−23
70.0	0.45	3 181	3 704	−524	−1 496	−83
80.0	0.41	3 314	4 234	−920	−2 628	−146
90.0	0.38	3 432	4 763	−1 331	−3 803	−211
100.0	0.35	3 538	5 292	−1 754	−5 012	−278
120.0	0.31	3 723	6 350	−2 628	−7 507	−417
150.0	0.26	3 953	7 938	−3 985	−11 386	−633
180.0	0.23	4 144	9 526	−5 382	−15 376	−854
210.0	0.21	4 309	11 113	−6 805	−19 442	−1 080
240.0	0.19	4 453	12 701	−8 247	−23 564	−1 309
300.0	0.16	4 701	15 876	−11 175	−31 929	−1 774
360.0	0.14	4 909	19 051	−14 143	−40 408	−2 245
420.0	0.12	5 089	22 226	−17 138	−48 965	−2 720
480.0	0.11	5 248	25 402	−20 154	−57 581	−3 199
540.0	0.10	5 392	28 577	−23 185	−66 243	−3 680
600.0	0.09	5 523	31 752	−26 229	−74 941	−4 163
660.0	0.09	5 643	34 927	−29 284	−83 669	−4 648
720.0	0.08	5 755	38 102	−32 348	−92 422	−5 135
840.0	0.07	5 957	44 453	−38 496	−109 989	−6 110
960.0	0.06	6 137	50 803	−44 667	−127 619	−7 090

* Needed Trench Length Matches Assumed Length

Figure 5.9 Example rational formula method for percolation trench sizing.

and they add to the treatment train as stormwater runoff migrates from its origin to the receiving waters.

Porous Pavement. When properly designed and operating, porous pavement can infiltrate, or otherwise treat, the runoff from 70 to 90% of all storm events. It, in effect, reduces the amount of directly connected impervious surface within a catchment.

There are three types of porous pavement: porous asphalt pavement, porous concrete pavement, and modular porous concrete block. Porous asphalt and concrete are constructed similarly to a conventional pavement, except sand and finer fraction of the aggregate are left out of the pavement mix. Such pavement typically is placed on top of a layer of granular base. The modular block pavement is constructed by placing the blocks over a layer of coarse gravel, which in turn is located on a porous geotextile fabric layer.

Whenever the porous pavement is expected to provide local disposal, the seasonal high groundwater and bedrock needs to be at least 0.9 m (3 ft) below the pavement's bottom. Typical porous pavement cross sections are illustrated in Figure 5.10. When the underlying soils, groundwater depth, or bedrock do not qualify the site for stormwater infiltration, porous pavement can be designed to be an underground detention facility. This can be done by installing an impermeable membrane between coarse rock media and the native soil subgrade. The granular base is then drained with the aid of perforated pipes installed at 3- to 8-m (10- to 25-ft) intervals. The release rate is controlled by a flow regulator, such as an orifice, which is designed to empty the pore storage volume within 6 to 12 hours.

Urbonas and Stahre (1993), after discussions with a number of public works departments, concluded that porous concrete and asphalt pavements have a tendency to seal and clog within 1 to 3 years. Also, faster sealing rates were reported in areas where extensive winter salting and sanding occur. One notable exception to surface clogging reports were the concrete pavement installations in the state of Florida (U.S.). Concrete and asphalt pavements need vigorous maintenance and the use of high-power vacuuming, though even then they seem to eventually seal. After sealed, this type of pavement has to be replaced.

Interlocking cellular concrete block pavement seems to seal at a slower rate than concrete and asphalt porous pavement and has a good record of service under a wide range of climatic conditions. After being sealed, the open spaces of modular blocks can be cleaned out by removing the vegetated soil or the sand layer and replacing it with fresh material. Individual blocks can settle and become misaligned, however, and this type of pavement is best suited for nontraffic areas such as parking pads and for overflow parking areas in sport event complexes, shopping centers, churches, and schools.

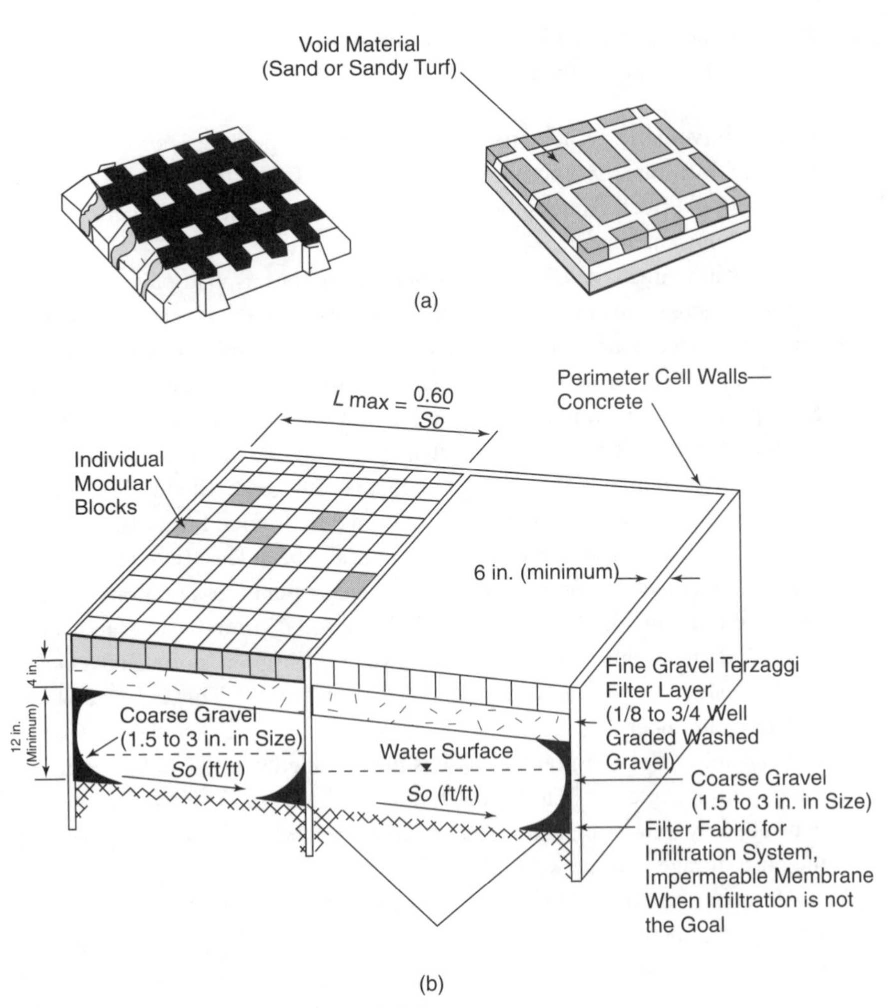

Figure 5.10 **Typical cross sections of porous pavement: (a) two examples of individual concrete modular paving block and (b) perspective of side-by-side modular block cells (ft × 0.304 8 = m; in. × 25.4 = mm).**

EXTENDED DETENTION (DRY) BASINS

Detention of urban stormwater runoff began to appear as an urban stormwater management practice in the early 1970s in North America, Europe, and

Australia to control runoff peaks from new land development sites. This was initially applied to control the 10-, 25-, 50-, or 100-year flow rates. By the mid- and late-1970s, Ontario (Canada) and the state of Maryland (U.S.) mandated detention to control the 2-year peak flow rate for stream bank erosion control purposes (with little success, as was determined later). The use of detention to control stormwater quality began to be used in the early 1980s. By the late 1980s, sufficient empirical data were available to design extended detention basins (that is, dry detention basins) for water quality purposes with reasonable confidence in their performance. Extended detention basins are best at removing suspended constituents. They are not particularly effective in removing solubles. Also, removal rates of solids by retention ponds tend to outperform detention basins. A comparison of constituent removal efficiencies of extended detention basins and retention ponds is described later in this chapter.

SIZING DETENTION BASINS. Using the Maximized Volume. There are several ways to size an extended detention basin. The simplest and most direct way for smaller catchments serving up to approximately 1.0 km^2 (0.6 sq mile) is to use the maximized volume described earlier in this chapter. The volume may be found for locations in the U.S. for basins emptying their entire volume in 24 and 48 hours. If one wishes to use another emptying time, simply interpolate between the results found using the 24- and 48-hour time. It is suggested that the event-capture-ratio–based coefficients in Table 5.4 be used with Equation 5.2 instead of the volume capture ratio coefficients.

The emptying, or drain, time is chosen by the designer or dictated by local authorities. Longer emptying times produce somewhat better removal rates of suspended solids. However, longer drain times tend to produce less attractive facilities, ones that have little or no vegetation on the bottom. Facilities with long emptying times have "boggy" bottoms with marshy vegetation and can be difficult to maintain and clean.

Using Hydrograph Routing. For detention basins that serve areas larger than 1.0 km^2 (0.6 sq mile), the volume can be found using a reservoir routing method. Again, it is recommended that the maximized storm depth be used. It has to be first converted to a design hyetograph, however, to simulate a runoff hydrograph. How this is done will be dictated by the typical design storm temporal distribution in use within the region where the facility is located. It is suggested, however, that the maximized depth be redistributed into a 2-hour design storm hyetograph.

The goal of reservoir routing is to balance inflow rates against outflow rates to find the needed volume. This is accomplished by solving Equation 5.9 with numerical methods or using one of the many available computer programs written for this purpose (see Chapter 3). Referring to Figure 5.11, Equation 5.10 states that the needed storage volume is a time integral of the difference between inflow and outflow hydrographs from the beginning of

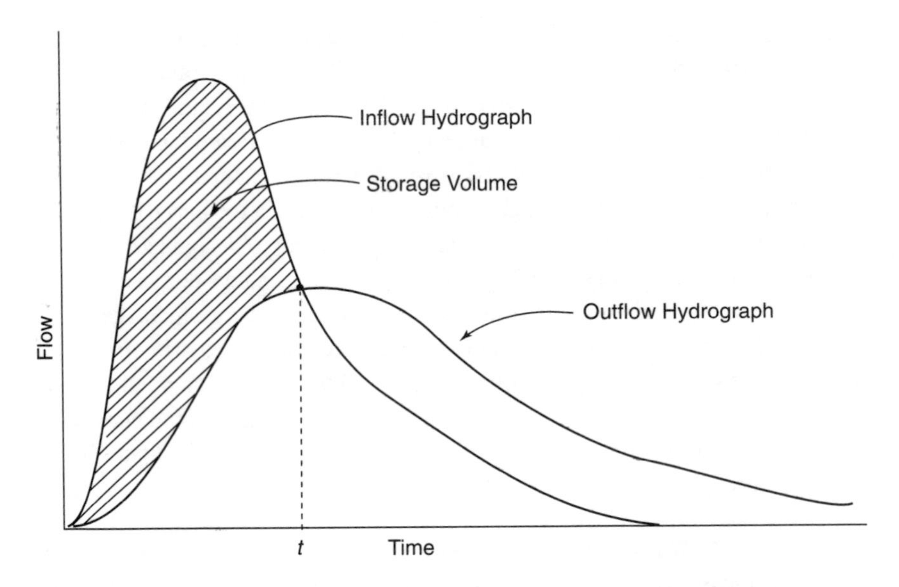

Figure 5.11 Routing a hydrograph through a detention basin.

storm runoff to the point in time where the outflow rate exceeds the inflow rate:

$$V_{max} = \int_0^t (Q_{in} - Q_{out})\, dt \tag{5.10}$$

Where

V_{max} = storage volume,
t = time from beginning of runoff to a point of maximum storage,
Q_{in} = Q_{out} on hydrograph recession limb,
Q_{in} = inflow rate, and
Q_{out} = outflow rate.

CONFIGURING AN EXTENDED DETENTION BASIN. In configuring an extended detention basin, try to make these facilities an integral part of the community. Consider multiple uses, aesthetics, safety, and the way the facility will fit into the urban landscape. Also, maintainability is an important consideration. Although these basins provide passive treatment with no operational attention, continued successful performance will depend on good maintenance. Always provide adequate maintenance access.

Figure 5.12 shows an idealized layout for an extended detention basin. The individuality of each on-site or regional facility and its place within the urban community make it incumbent on the designer to seek out local input, identify site constraints, identify the community's concerns, and consider a wide array of possibilities during design.

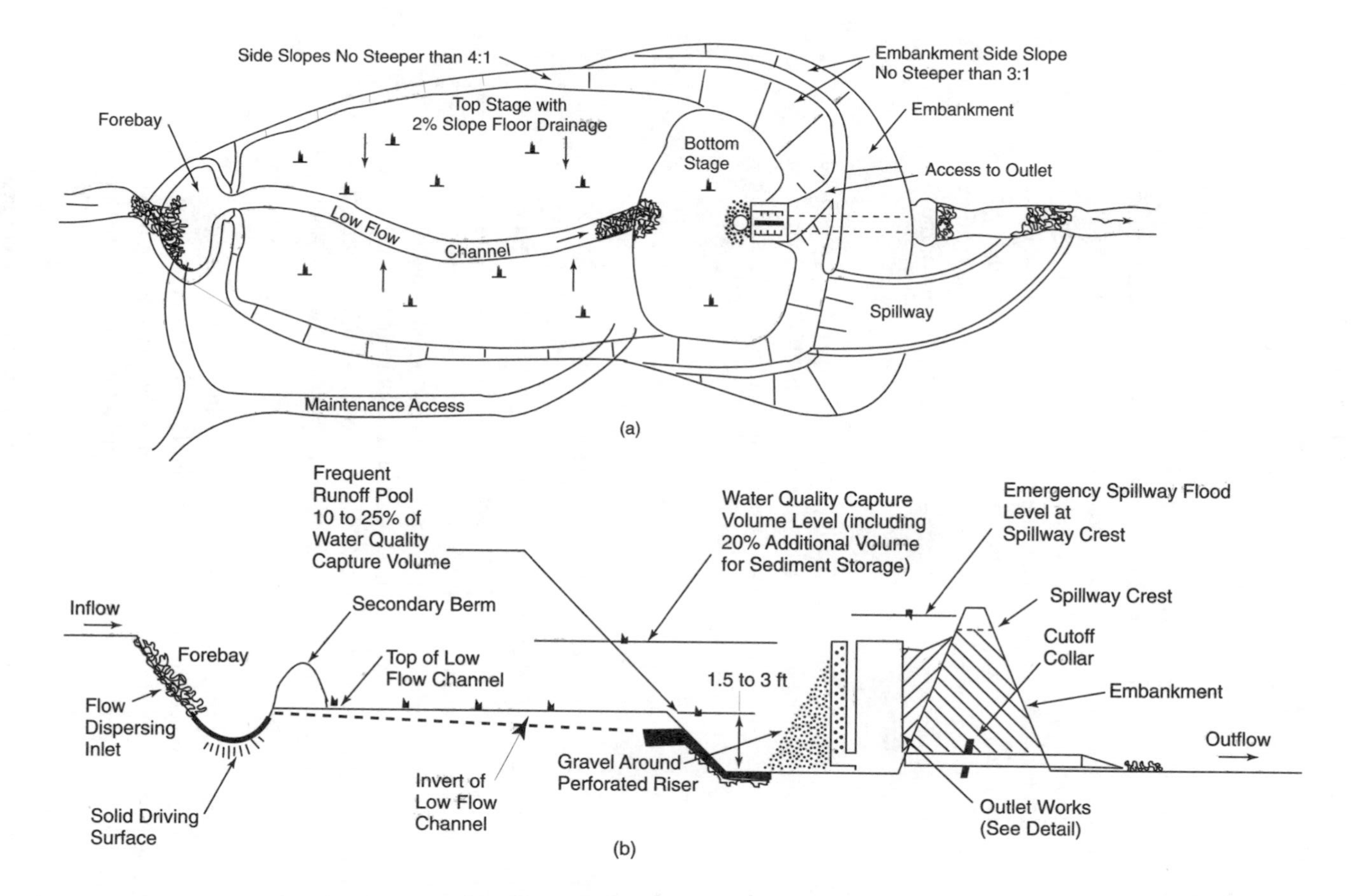

Figure 5.12 **An idealized extended detention basin: (a) plan (not to scale) and (b) section (not to scale) (ft × 0.304 8 = m) (UDFCD, 1992).**

Storage Volume. Provide a storage volume, sometimes called capture volume, equal to the maximized volume described earlier. Add 20% to this volume to provide for sediment accumulation. Randall *et al.* (1982) and Whipple and Hunter (1981) suggest that such detention basins be designed to promote sedimentation of small particles, namely smaller than 60 microns in size, which account for approximately 80% of the suspended sediment mass found in stormwater (Urbonas and Stahre, 1993).

Provide an outlet to empty less than 50% of the design volume in the first one-third of the design emptying period (that is, 12 to 16 hours). This ensures that small runoff events will be detained to remove small suspended solids. The particles, as they settle in a water column, concentrate at the lower levels of the temporary pond. It is this layer that drains through the bottom outlet of a detention basin. A long emptying time–thus the term extended detention–permits smaller particles to attach to the bottom of the basin and become trapped.

Flood Control Storage. Whenever feasible, try to incorporate the extended detention basin within a larger flood control facility. The designer may want to consider combining water quality and flood control functions in a single detention basin.

Basin Geometry. The basin should gradually expand from the inlet and contract toward the outlet to reduce short circuiting. Provide a length-to-width ratio of two or greater, preferably up to a ratio of four.

Two-Stage Design. Whenever feasible, provide a two-stage basin. The lower portion has a micropool that fills often. This reduces the periods of standing water and sediment deposition in the remainder of the basin. The top stage should be 0.6 to 1.8 m (2 to 6 ft) deep, its bottom sloping at approximately 2% toward a low-flow channel. The bottom pool can be 0.5 to 0.9 m (1.5 to 3 ft) deeper and should be able to store 15 to 25% of the capture volume. These recommendations do not necessarily apply to large, regional extended detention basins.

Basin Side Slopes. Basin side slopes need to be stable under saturated soil conditions. They also need to be sufficiently gentle to limit rill erosion, facilitate maintenance, and address the safety issue of individuals falling in when the basin is full of water. Side slopes of 4:1 and flatter provide well for these concerns.

Forebay. Design the basin to encourage sediment deposition to occur near the point of inflow. A forebay with a volume equal to approximately 10% of the total design volume can help with the maintenance of the basin, and the service life of the remainder of the basin can be extended. Equip it with a stabilized access and a concrete or soil cement lined bottom to prevent mechanical equipment from sinking to the bottom.

Basin Inlet. Most erosion and sediment deposition occurs near the inlet. An ideal inflow structure will convey stormwater to the basin while preventing erosion of the basin's bottom and banks, reducing resuspension of previously deposited sediment and facilitating deposition of heaviest sediment near the inlet. With several compromises, many of these design goals can be nearly achieved. Inflow structures can be drop manholes, rundown chutes with an energy dissipator near the bottom, a baffle chute, a pipe with an impact basin, or one of the many other types of diffusing devices.

Low-Flow Channel. Provide a low-flow channel to convey trickle flows and the last of the captured volume to the outlet.

Outlet Design. Use an outlet capable of slowly releasing the design capture volume over the design emptying time. One example is a perforated riser, illustrated in Figure 5.13. Another arrangement of an outlet was suggested by Schueler *et al.* (1992), namely, a hooded perforated riser located in a small permanent pool (that is, a micropool).

Because extended detention basins are designed to encourage sediment deposition and urban stormwater has substantial quantities of settleable and floatable solids, basin outlets are prone to being clogged. This can make the design of reliable outlet structures for extended detention basins difficult. A clogged outlet will invalidate the hydraulic function of even the best design.

ASCE (1985), ASCE (1992), DeGroot (1982), Roesner *et al.* (1989), Schueler (1987), Schueler *et al.* (1992), Urbonas and Roesner (Eds.) (1986), and Urbonas and Stahre (1993) reported many reasons for outlet problems, which include clogging by trash and debris, silting in of the outlet, damage by vandalism, children plugging an outlet, and other factors that modify its discharge characteristics. Each outlet has to be designed with clogging, vandalism, maintenance, aesthetics, and safety in mind.

Trash Rack. If the outlet is not protected by a gravel pack, as shown in Figure 5.13, provide some form of a trash rack. Never wrap a perforated outlet in a geotextile filter cloth that will seal quickly. Figure 5.14 is a chart that provides simple, empirically based guidance for minimum sizes of trash racks for detention outlets.

Dam Embankment. Design and build the dam embankment so that it will not fail during storms larger than the water quality design storm. Provide an emergency spillway or design the embankment to withstand overtopping commensurate with the size of the embankment, the volume of water that can be stored behind it, and the potential of downstream damages or loss of life if the embankment fails. Emergency spillway designs vary widely with local regulations. Embankments for small on-site basins should be protected from at least the 100-year flood, while the larger facilities should be evaluated for the probable maximum flood. Always consult the state's dam regulatory agency.

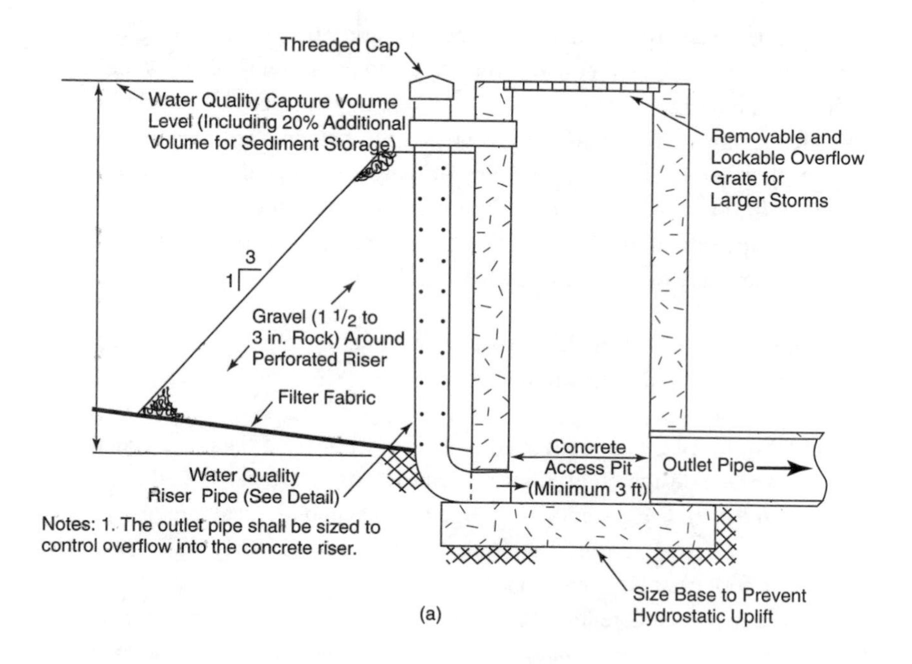

Notes 1. Minimum number of holes = 8.
2. Minimum hole diameter = 1/8 in. diameter.

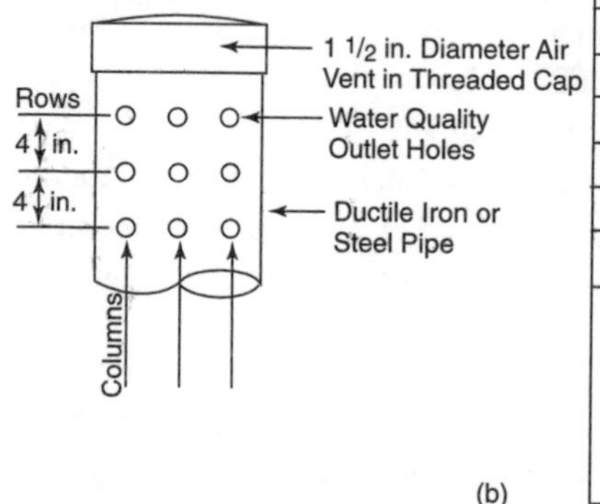

Maximum Number of Perforated Columns				
Riser Diameter, in.	Hole Diameter, in.			
	1/4 in.	1/2 in.	3/4 in.	1 in.
4	8	8	±	±
6	12	12	9	±
8	16	16	12	8
10	20	20	14	10
12	24	24	18	12

Hole Diameter, in.	Area of Hole, sq in.
1/8	0.013
1/4	0.049
3/8	0.110
1/2	0.196
5/8	0.307
3/4	0.442
7/8	0.601
1	0.785

(b)

Figure 5.13 **An example of a perforated riser outlet: (a) outlet works (not to scale) and (b) water quality riser pipe (not to scale) (ft × 0.304 8 = m; in. × 25.4 = mm) (UDFCD, 1992).**

Embankment slopes should be no steeper than 3:1, preferably 4:1 or flatter. They also need to be planted with turf-forming grasses. Embankment soils should be compacted to 95% of their maximum density at optimum moisture.

Vegetation. A basin's vegetation provides erosion control and enhances sediment entrapment. The basin can be planted with native grasses or with ir-

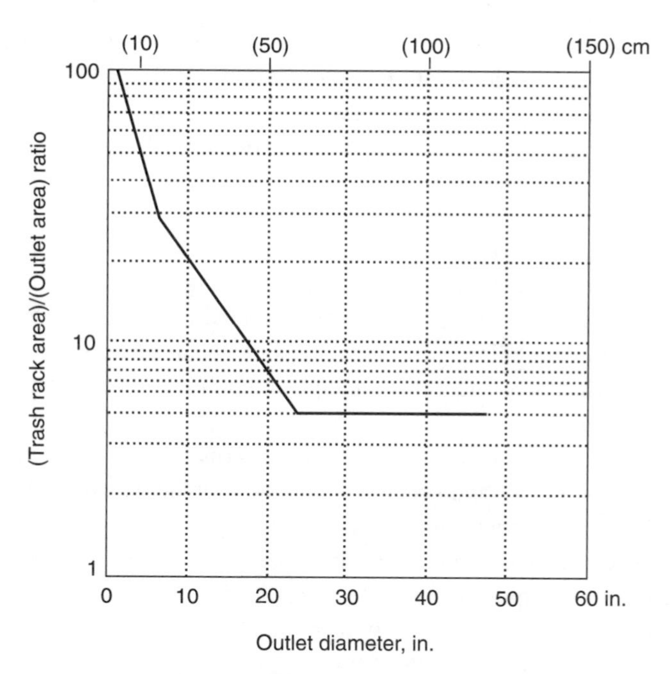

Figure 5.14 **Minimum size of a trash rack versus outlet diameter (note: trash rack area is the net area of all openings between bars, rock packing, and so on) (in. × 25.4 = mm) (UDFCD, 1992).**

rigated turf, depending on the local setting, basin design, and its intended other uses (such as recreation). Sediment deposition, along with frequent and prolonged periods of inundation, make it difficult to maintain healthy grass cover on the basin's bottom. Options for an alternative bottom liner include a marshy wetland bottom, bog, layer of gravel, riparian shrub, bare soil, low-weed species, or other type that can survive the conditions found on the bottom of the basin.

Maintenance Access. Provide for vehicular maintenance access to the forebay and the outlet areas with grades that do not exceed 8 to 10% and have a stable surface of gravel-stabilized turf, a layer of rock, or concrete pavement.

Multiple Uses. Whenever desirable and feasible, incorporate the water quality basin within a larger flood control facility. Also, whenever possible, provide for other uses such as active or passive recreation, wildlife habitat, or wetland. The use of a multiple-stage basin design described earlier can help accommodate multiple uses. The area within an extended detention basin is not well suited for active recreation such as playing fields. These are best located above an extended detention basin's pool level.

Aesthetics. Aesthetics are what the public uses to judge how "successful" or "useful" a detention basin is within the community. Although there are examples of unattractive basins, most new facilities are tastefully integrated to the neighborhood. Aesthetics are important. Using a landscape architect to assist with the design should be considered.

Safety. For larger on-site basins and regional facilities, safety has to also include the structural integrity of the water impounding embankment. As discussed earlier, the embankment should be protected from catastrophic failure. In the U.S., dam failure is almost always judged as an absolute liability of its owner. Always consider this principle of common law when designing detention facilities.

When the facility is in operation, safety concerns need to focus on flow velocities, water depths, and keeping the public from being exposed to high-hazard areas. During dry weather periods, safety is enhanced by reducing the use of high vertical walls and steep side slopes. Outlets and inflow structures and adjacent areas require special attention, and ASCE (1985) suggests the use of thorny shrubs and trash/safety racks at all outlet orifices, pipes, and weirs.

SUMMARY AND CONCLUSIONS. Extended detention basins are viable and effective treatment facilities. When properly designed, reductions of approximately 70% are possible in the total suspended sediment load and of constituents associated with these sediments. Regional facilities often offer economies of scale and greater reliability in capturing stormwater when they are used, while on-site facilities offer institutional and fiscal advantages of implementation as the land is urbanized.

RETENTION PONDS (WET)

A retention pond is a small artificial lake with emergent wetland vegetation around the perimeter, designed to remove pollutants from stormwater. This BMP is also sometimes called a "wet pond" or a "wet detention basin." This manual refers to it as a retention pond to distinguish it from the extended detention basin described in the previous section. A retention pond often is sized to remove nutrients and dissolved constituents, while any pool that may be associated with an extended detention basin is smaller and is provided for aesthetics, for example, to cover solids-settling areas.

Features of a retention pond are shown in Figure 5.15. The permanent pool provides a vessel for the settling of solids between storms and the removal of nutrients and dissolved pollutants. The wetland vegetation bench, called the littoral zone, provides aquatic habitat, enhances pollutant removal, and reduces the formation of algal mats. Figure 5.15 also shows an optional surcharge detention storage volume overlying the permanent pool. This can

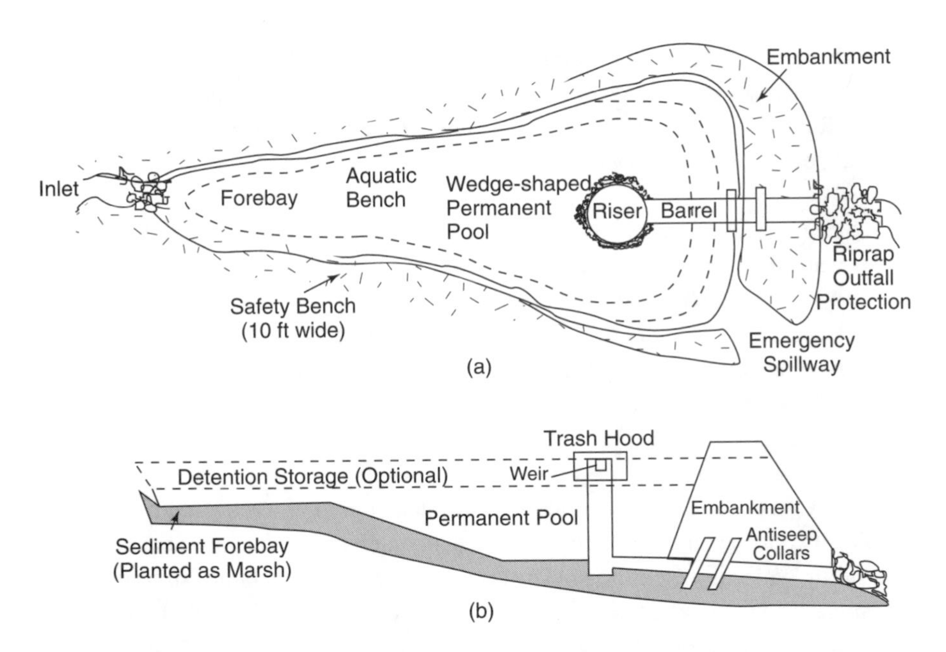

Figure 5.15 Plan and profile of a retention basin: (a) top view and (b) side view (Schueler, 1987).

be used for flood control. Some local jurisdictions require that this surcharge storage be designed as extended detention for added pollutant removal efficiency.

STORMWATER MANAGEMENT APPLICATIONS. Control of Nutrient Loadings. Retention ponds can be superior to extended detention basins for the control of nutrients in urban stormwater. While detention basins rely on solids-settling processes, retention ponds remove dissolved nutrients through several physical, chemical, and biological processes in the permanent pool. Table 5.13 shows a comparison of removal efficiencies of

Table 5.13 Comparison of pollutant removal percentages by well-designed extended detention basins and retention ponds (U.S. EPA, 1983).

Type of practice	Total suspended sediments	Nitrogen	Phosphorus	Lead	Zinc	Biochemical oxygen demand
Extended detention	70–80	0 (Diss) 20–30 (Total)	0 (Diss) 20–50 (Total)	70–80	40–50	20–40
Retention ponds	70–80	50–70 (Diss) 30–40 (Total)	50–70 (Diss) 50–60 (Total)	70–80	40–50	20–40

properly sized retention ponds and extended detention basins. In addition, petroleum hydrocarbon removals are similar to those of total suspended sediments. Retention ponds are most appropriate where nutrient loadings are of concern, especially in the following situations:

- Watersheds tributary to reservoirs and lakes—retention ponds in the watershed can help achieve eutrophication management goals in downstream reservoirs and lakes.
- Watersheds tributary to tidal embayments and estuaries—nutrient loadings into estuarine systems is a growing concern in coastal areas, including upland areas that drain into tidal waters. Retention ponds can help reduce the nutrient loads.

Removal of nutrients has a price: the permanent pool of a retention pond requires two to seven times more volume than an extended detention basin, depending on local meteorology. The larger volume requires larger structures and more land than detention basins, resulting in costs of facilities that are 50 to 150% more than for extended detention basins. If, however, the facility requires overlying storage for flood control peak-shaving, cost increases become smaller as the flood control volume and benefits get larger. Table 5.14, which summarizes design criteria for a regional stormwater management master plan for Fairfax County, Virginia (U.S.), exemplifies the relative difference in size for retention ponds and extended detention basins for this region of the U.S.

Table 5.14 Comparison of detention storage requirements in Fairfax County, Virginia: permanent pool of retention pond versus extended detention basin.[a]

Land use	Imperviousness, %	Retention pond, in.[b]	Extended detention,[c] in.[b]
Low-density single family	20	0.7	0.1
Medium-density single family	35	0.8	0.2
Multifamily residential	50	1.0	0.4
Industrial/office	70	1.2	0.5
Commercial	80–90	1.3	0.8
Forest/undeveloped	0	0.5	0.0

[a] Retention pond pool volume is based on an average hydraulic retention time of 2 weeks.
[b] In. × 25.40 = mm.
[c] Extended detention volume is based on the capture of first-flush runoff.

Aesthetics. Retention ponds offer a number of aesthetic advantages. They typically are more attractive than extended detention basins and are considered property value amenities in many areas. This is because sediment and debris accumulate within the permanent pool and are out of sight.

Other Siting Considerations. Retention ponds can be designed to require little hydraulic head to operate. While Figure 5.15 shows a dam at the downstream end of the pond, in flat terrain the permanent pool can be excavated below the ground surface, a common practice in the state of Florida (U.S.). Before excavating into the groundwater table, check with local regulatory authorities; however, if the pond is not sited over a gravel or karst formation, it should not adversely affect the quality of the groundwater, although such a possibility may exist in some cases. Most pollutants typically are removed from the groundwater in the first 0.4 to 0.9 m (18 to 36 in.) of soil downgradient of the pond.

Other issues to consider when choosing a retention pond include

- If the tributary catchment is large enough to have sufficient base flow to sustain a permanent pool;
- If the receiving waters immediately downstream are particularly sensitive to increased effluent water temperatures that can result from introduction of the pond;
- If existing wetlands at the site restrict the use of a permanent pool; and
- If water rights available for evapotranspiration are consumptive use in states with a prior appropriation water law system.

DESIGN METHODS. Two different methods are used for the design of permanent pools for a retention pond:

- Solids-settling design method relies on the solids-settling theory and assumes that all pollutant removal is because of sedimentation (Driscoll, 1983, and U.S. EPA, 1983).
- Lake eutrophication model design method provides for a level of eutrophication by accounting for the principal nutrient removal mechanisms (Hartigan, 1989, and Walker, 1987).

Solids-Settling Design Method. The solids-settling method is most appropriate for situations where the control of total suspended sediments and pollutants that attach themselves to the solids is the principal objective. The method relies on rainfall and runoff statistics, pond size, and settling velocities of suspended solid particle size distributions to calculate total suspended sediment removal. This method assumes an approximate plug flow system in the retention pond with all pollutant removal resulting from sedimentation.

Retention pond design curves based on the solids-settling method are shown in Figure 5.16 (U.S. EPA, 1986) for low-density, single-family resi-

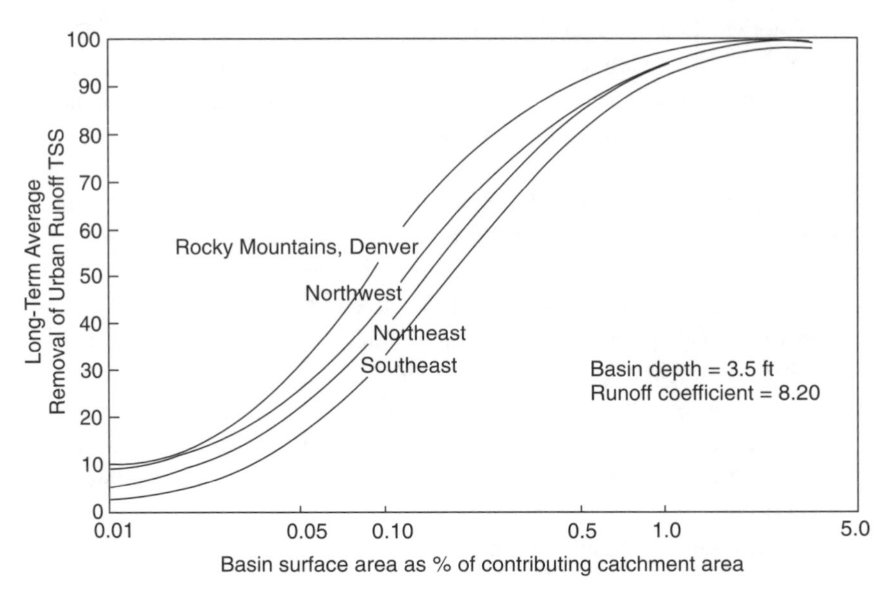

Figure 5.16 Geographically based design curves for solids settling model (ft × 0.304 8 = m) (U.S. EPA, 1986).

dential development (runoff coefficient, $RV = 0.2$). Separate design curves must be developed for other land-use patterns. These design curves relate average total suspended sediment removal to the size of the permanent pool. Here, the permanent pool size is expressed as the ratio of its surface area to the tributary catchment area and is based on a mean pool depth of approximately 1 m (3.5 ft). Average removal rates for other constituents may be estimated by multiplying the total suspended sediment removal rate by the average particulate fraction of the constituent of interest.

The total suspended sediment settling model was tested using data from nine retention ponds monitored during U.S. EPA's National Urban Runoff Program (NURP). Based on goodness-of-fit plots, it was concluded that the method did a reasonably good job of predicting removal rates at these nine NURP sites (Driscoll, 1983, and U.S. EPA, 1986).

A drawback to the performance curves shown in Figure 5.16 is that they were developed for only low-density residential land use (that is 20 to 30% imperviousness). However, comparison of this method with facilities designed using the maximized volume based on a 12-hour emptying time has shown nearly identical results, provided a surcharge extended detention volume is given. Thus, it is recommended that the permanent pool volume for the solids-settling method be the maximized volume based on a 12-hour drain time as described earlier in this chapter. It is also recommended that a surcharge extended detention volume be provided above the permanent pool, which is also equal to the maximized runoff volume. The outlet is then designed to draw down, or empty, this surcharge volume in 12 hours. This type of design significantly improves pollutant removal efficiency and virtually eliminates many of the short-circuiting problems often found in ponds without a surcharge extended detention volume above the permanent pool.

Lake Eutrophication Model Design Method. This method assumes that a retention pond is a small eutrophic lake that can be represented by empirical models used to evaluate lake eutrophication effects (Hartigan, 1989, and Walker, 1987). Using this design method, a retention pond can be sized to achieve a controlled rate of eutrophication and an associated removal rate for nutrients. Because retention ponds that achieve nutrient removal also remove other pollutants, typically it is not necessary for the design process to address constituents other than nutrients. Also, as noted earlier, large retention ponds may not be cost effective unless nutrient control is the principal water quality management objective.

The lake eutrophication design model is the phosphorus retention coefficient model developed by Walker (1985 and 1987). Like most input/output lake eutrophication models, this model is an empirical approach that treats the permanent pool as a completely mixed system and assumes that it is not necessary to consider the temporal variability associated with individual storm events. Unlike the solids-settling model, which accounts for temporal variability of individual storms, the Walker model is based on annual flows and loadings.

The model is applied in two parts:

$$K_2 = \frac{0.56 \cdot Q_S}{F \cdot (Q_S + 13.3)} \tag{5.11}$$

Where

K_2 = second-order decay rate, $m^3/mg{\cdot}a$;
Q_S = Z/T the mean overflow rate, m/a;
Z = mean pond depth, m;
T = average hydraulic retention time, years; and
F = inflow (ortho P)/(total P) ratio.

$$R = 1.0t \frac{1.0 - \sqrt{1.0 + (1.0 + 4N)}}{2N} \tag{5.12}$$

Where

R = total P retention coefficient (that is, BMP efficiency),
N = $K_2 \cdot P_T \cdot T$, and
P_T = inflow total P, $\mu g/L$.

Equations 5.11 and 5.12 were developed from a database for 60 U.S. Army Corps of Engineers' reservoirs and were verified for 20 other reservoirs. The model was applied by Walker (1987) to 10 NURP sites and 14 other retention pond systems and small lakes. The goodness-of-fit test yielded an $R^2 = 0.8$, indicating a good job of replicating monitored total P removals.

The permanent pool storage volume, V_B, is calculated for the desired average removal rate for P_t, which is a function of the average hydraulic retention time, T. The value of T (in years) is computed by dividing the permanent pool volume, V_B, by the product of the mean storm runoff, V_R, times the total number of runoff events per year, n, namely, $T = V_B/(V_R \cdot n)$. Field studies indicate that an optimum removal rate for T_P of approximately 50% occurs at T values of 2 to 3 weeks for pools with mean depths of 1.0 to 2.0 m (3 to 6 ft) (Hartigan, 1989). In the eastern U.S., this optimum range for T values corresponds to V_B/V_R ratios of 4 to 6. Ponds with values of T greater than 2 to 3 weeks have a greater risk of thermal stratification and anaerobic bottom waters, resulting in an increased risk of significant export of nutrients from bottom sediments.

State and regional stormwater management regulations and guidelines often address design criteria for the permanent pool storage volume in terms of either average hydraulic retention time, T, V_B/V_R, or minimum total suspended sediment removal rate. For example, the U.S. state of Florida (Florida DER, 1988) requires an average hydraulic retention time of 14 days, equivalent to V_B/V_R of 4; the Urban Drainage and Flood Control District's BMP criteria manual in the Denver, Colorado, area (U.S.) (UDFCD, 1992) specifies that the permanent pool storage volume should be 1.0 to 1.5 times the "water quality capture volume," which is equivalent to V_B/V_R on the order of 1.5 to 2.5. A municipal BMP handbook published by the California State Water Resources Control Board (Camp Dresser & McKee *et al.,* 1993) recommends that retention pond permanent pools be sized for a V_B/V_R of 3. The U.S. state of North Carolina's stormwater disposal regulations for coastal areas and water supply watersheds specify that the permanent pool should be sized to achieve a total suspended sediment removal rate of 85%, which is equivalent to a V_B/V_R in the range of 3 to 4 when no surcharge extended detention is provided. With surcharge extended detention, 85% removal of total suspended sediments has been achieved with a V_B/V_R of 2 or less.

CONFIGURING A RETENTION POND. Depth of Permanent Pool. Mean depth of the permanent pool is calculated by dividing the storage volume by the surface area. The mean depth should be shallow enough to ensure aerobic conditions and reduce the risk of thermal stratification but deep enough to ensure that algal blooms are not excessive and reduce resuspension of settled pollutants during significant storm events. The minimum depth of the open water area should be greater than the depth of sunlight penetration to prevent emergent plant growth in this area, namely, on the order of 2 to 2.5 m (6 to 8 ft).

A mean depth of approximately 1 to 3 m (3 to 10 ft) should produce a pond with sufficient surface area to promote algae photosynthesis and should maintain an acceptable environment within the permanent pool for the average hydraulic retention times recommended above, although separate analyses should be performed for each locale. If the pond has more than 0.8 ha (2 ac) of water surface, mean depths of 2 m (6.5 ft) will protect it against wind-generated resuspension of sediments. The mean depths of the more effective retention ponds monitored by the NURP study typically fall within this range. A water depth of approximately 1.8 m (6 ft) over the major portion of the pond will also increase winter survival of fish (Schueler, 1987).

A maximum depth of 3 to 4 m (10 to 13 ft) should reduce the risk of thermal stratification (Mills *et al.*, 1982). However, in the U.S. state of Florida, pools up to 9.2 m (30 ft) deep have been successful when excavated in high groundwater areas; this is probably because of improved circulation at the bottom of the pond as a result of groundwater moving through it.

Side Slopes along Shoreline and Vegetation. Side slopes along the shoreline of the retention pond should be 4H:1V or flatter to facilitate maintenance (such as mowing) and reduce public risk of slipping and falling into the water. In addition, a littoral zone should be established around the perimeter of the permanent pool to promote the growth of emergent vegetation along the shoreline and deter individuals from wading (see Figure 5.15). The emergent vegetation around the perimeter serves several other functions: it reduces erosion, enhances the removal of dissolved nutrients in urban stormwater discharges, may reduce the formation of floating algal mats, and provides habitat for aquatic life and wetland wildlife. This bench for emergent wetland vegetation should be at least 3 m (10 ft) wide with a water depth of 0.15 to 0.45 m (0.5 to 1.5 ft). The total area of the aquatic bench should be 25 to 50% of the permanent pool's water surface area. Local agricultural agencies or commercial nurseries should be consulted about guidelines for using wetland vegetation within shallow sections of the permanent pool.

Extended Detention Zone above the Permanent Pool. Some state or local regulations require detention of a specified runoff volume as surcharge above the permanent pool. Storage in the surcharge zone is released during a specified period through an outlet structure. This surcharge detention requirement is intended to reduce short circuiting and enhance settling of total suspended sediments. Settling-solids analysis shows that retention ponds sized for nutrient removal with a minimum detention time, T, of 2 weeks and a minimum V_B/V_R of 4 achieve total suspended sediment removal rates of 80 to 90%. Addition of an extended detention zone above the permanent pool is unlikely to produce measurable increases in the removal of total suspended sediments. Still, a surcharge extended detention volume is recommended whenever the V_B/V_R is less than 2.5. Whenever one is used or required, it is suggested that the maximized event-based volume with a 12-hour drain time be used.

Minimum and Maximum Tributary Catchment Areas. The minimum drainage area should permit sufficient base flow to prevent excessive retention times or severe drawdown of the permanent pool during dry seasons.

Unless regional experience is available for determining the minimum drainage area required in a particular location, it is recommended that a water balance calculation be performed using local runoff, evapotranspiration, exfiltration, and base flow data to ensure that the base flow is adequate to keep the pond full during the dry season.

The maximum tributary catchment area should be set to reduce the exposure of upstream channels to erosive stormwater flows, reduce effects on perennial streams and wetlands, and reduce public safety hazards associated with dam height. Again, regional experience will be useful in providing guidelines. For example, in the southeastern U.S., some stormwater master plans have restricted the maximum tributary catchments to 40 to 120 ha (100 to 300 ac) depending on the amount of imperviousness in the watershed, with highly impervious catchments restricted to the lower end of this range and vice versa. On the other hand, experience in semiarid areas has shown that even a small area of new land development can cause downstream erosion and that drainageway stabilization is needed between the new development and the pond for relatively small catchments.

Construction of Retention Ponds in Wetland Areas. One potential constraint on the use of retention ponds as regional BMPs is federal regulations that restrict the filling of wetland areas and the Section 404 permit program regulating any wetland or retention pond constructed for stormwater management. Although retention pond BMPs typically are designed to enhance pollutant removal by incorporating wetland areas along the perimeter, regulatory agencies may restrict their use if a significant amount of native wetlands will be submerged within the permanent pool. In addition, restorative maintenance of the created wetland areas, which includes removal of silt, may require a Section 404 permit. If work is performed without such a permit, the owner can be subject to federal and state enforcement action and fines. Thus, it is important to check with the local offices of the federal regulatory agencies, such as the U.S. Army Corps of Engineers and state regulators, about the need for such permits. A written response should always be obtained before proceeding with any restorative maintenance work.

Potential wetlands constraints must be addressed on a case-by-case basis during final design of each retention pond facility. If field inspections indicate that a significant wetlands area will be affected at a particular site, the following options can be pursued during final design:

- Investigate moving the embankment and permanent pool upstream of the major wetland area.
- If the above option is unfeasible, a wetland mitigation plan can be developed as a part of the retention pond design.
- If neither of the above options result in a design acceptable to regulatory agencies, consider using an extended detention basin instead. Eliminating the permanent pool can often reduce adverse effects on

native wetlands, but their direct oversight by regulatory agencies may not be avoided.

Basin Geometry. Relatively large length-to-width ratios can help reduce short circuiting, enhance sedimentation, and help prevent vertical stratification within the permanent pool. A minimum length-to-width ratio of 2:1 (3:1 preferred) is recommended for the permanent pool. The permanent pool should expand gradually from the basin inlet and contract gradually toward the outlet, maximizing the travel time from the inlet to the outlet. Baffles or islands within the pool can increase the flow path length and reduce short circuiting.

Soil Permeability. Highly permeable soils may not be acceptable for retention ponds because of excessive drawdown during dry periods. Where permeable soils are encountered, exfiltration rates can be minimized by scarifying and compacting a 0.3-m (12-in.) layer of the bottom soil of the pond, incorporating clay to the soil, or providing an artificial liner. Excavating the permanent pool into the groundwater table can also ensure its permanency, but seasonal fluctuations in the groundwater table need to be taken into account.

Forebay. To reduce the frequency of major cleanout activities within the pool area, a sediment forebay with a hardened bottom should be constructed near the inlet to trap coarse sediment particles. The forebay storage capacity should be approximately 10% of the permanent pool storage. Access for mechanized equipment should be provided to facilitate removal of sediment. The forebay can be separated from the remainder of the permanent pool by one of several means: a lateral sill with wetland vegetation, two ponds in series, differential pool depth, rock-filled gabions, a retaining wall, or a horizontal rock filter placed laterally across the permanent pool.

Inlet and Outlet Structures. The inlet design should dissipate flow energy and diffuse the inflow plume where it enters the forebay or permanent pool. Examples of inlet designs include drop manholes, energy dissipators at the bottom of paved rundown, a lateral bench with wetland vegetation, and the placement of large rock deflectors.

An outlet for a retention pond typically consists of a riser with a hood or trash rack to prevent clogging and an adequate antivortex device for basins serving large drainage areas. Some typical outlet structures are illustrated in Figure 5.17. Antiseep collars should be installed along outlet conduits passing through or under the dam embankment. If the pond is a part of a larger peak-shaving detention basin, the outlet should be designed for the desired flood control performance. An emergency spillway must be provided and designed using accepted engineering practices to protect the basin's embankment. Be certain that the pond embankment and spillway are designed in accordance with federal, state, and local dam safety criteria.

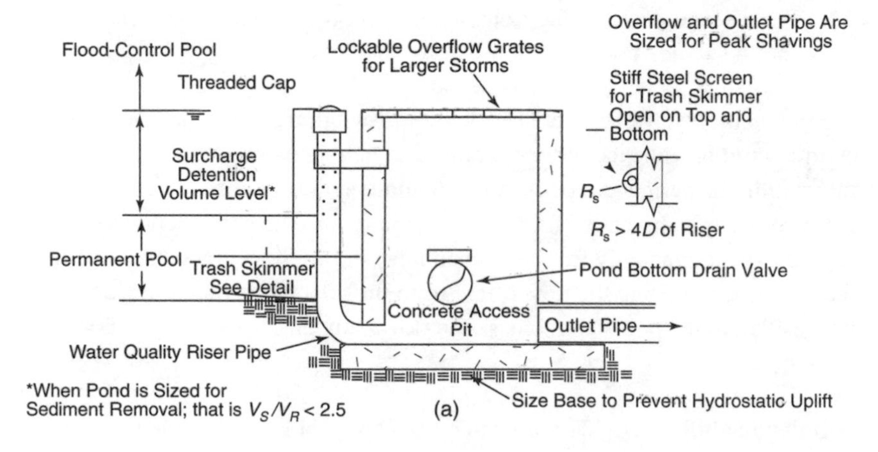

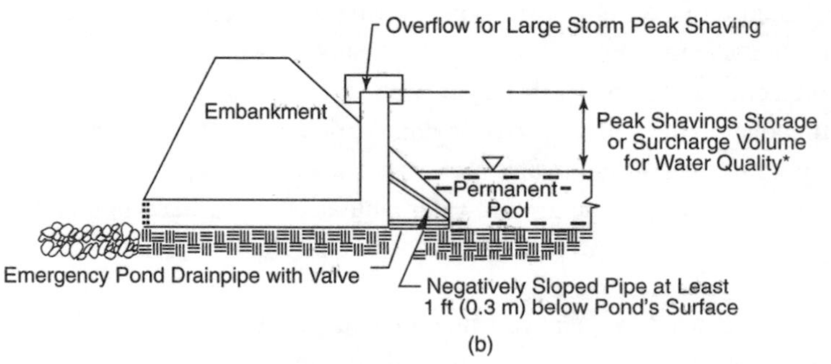

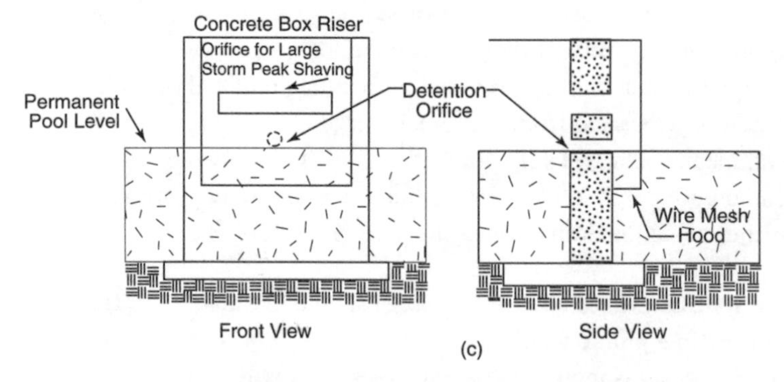

Figure 5.17 Typical outlet structures: (a) outlet works with surcharge detention for water quality, (b) negatively sloped pipe outlet with riser, and (c) multiple orifice outlet (Schueler, 1987, and UDFCD, 1992).

The channel that receives the discharge from the basin's outlet should be protected from erosive discharge velocities. Options include riprap lining of the channel or the provision of stilling basins, check dams, rock deflectors, or other devices to reduce outfall discharge velocities to nonerosive levels.

CONSTRUCTED WETLANDS

The use of constructed wetlands is popular for treating stormwater. Reported removal efficiencies vary. Strecker *et al.* (1992) summarized the performance of several wetlands in the U.S. for treatment of urban runoff; they found that suspended solids removals averaged 87%, with a range of 40 to 96%.

A significant deterrent to the comparison of removal efficiencies between wetlands is the lack of a standard set of design criteria. What is reported here is a general guidance based on information available in the literature.

DESIGN. General Considerations. Specific site conditions are important to the proper design of a wetland. Key site characteristics include soils, hydroperiod, and plant species and density. Depth to the confining layer or groundwater is important to ensure that the wetland does not dry up during extended periods of no rainfall. In addition, a constant source of surface water is recommended; stagnant water in the wetlands causes the underlying soil to become anaerobic, releasing ammonia, phosphorus, and heavy metals to the overlying water for washout during the next runoff event. Stagnant water also results in mosquito problems. Finally, the depth and duration of maximum submergence are important because an excess of either will kill the vegetation.

Hydraulic Design. The following hydraulic design criteria are recommended for wetlands:

- Maintain dry weather flow depths that vary through the wetland between 0.1 and 1.2 m (0.5 to 4 ft), depending on the types of vegetation planted, with the outlet structure designed so that the wetland can be periodically drawn down completely to dry the sediments (provides for natural oxidation of built-up organics);
- Size the wet weather storage volume using the methodology for extended detention basins but with a maximum surcharge depth above the dry weather flow depth of 0.6 m (2 ft) and a drawdown time of 24 hours; this will reduce stress on herbaceous wetland plants. The 0.6-m depth limitation will determine the surface area required for the wetland.
- Design inlet structures to achieve sheet flow across the wetland to the maximum extent possible.
- Design the outlet structure to control the water surface and protect it from plugging by floatables common in wetlands (see Inlet and Outlet Structures in this chapter).
- If open water is to be included in the wetland, it should be less than 50% of the total wetland area; the depth of the open water should follow the rules for the maximum permanent pool depth in retention ponds.

Configuration of the Wetland. The siting and configuration of the created wetland somewhat depend on adjacent land uses, the magnitude of contributing surface runoff, and the type of collection system (that is, shallow ditches or underground piping). Variations in topography and plant types will create more suitable habitat for wildlife. If the proposed site is large enough, some upland areas (peninsulas or islands) are preferable. Upland buffers increase the habitat value of a created wetland.

Figure 5.18 shows an idealized wetland basin designed to enhance stormwater quality. Urbonas and Stahre (1993) propose that the ideal shape is similar to an oval, with the outlet and inlet at opposite ends. If an oval shape is not possible, use any other elongated shape that separates the inlet and outlet as much as possible. The primary goals are to increase the contact time of the inflow with the wetland surfaces and ensure that the inflow does not short-circuit the facility. An elongated basin shape helps to achieve these goals. It is suggested that the length-to-width ratio of the wetland surface be no less than 3 (that is, $L/W > 3$); a 2:1 ratio is recommended by Livingston (1989).

The forebay shown in Figure 5.18 helps settle out the largest sediment particles before the flow passes over the areas covered with emergent vegetation. It also helps to spread the inflow uniformly over the entire wetland. The forebay should also have a baffle near the inlet as illustrated to break up the inflow jet and facilitate spreading of the inflow over the entire surface area of the wetland.

An overflow outlet, not dissimilar to a riser used in a retention pond, is placed within a deepened portion near the outlet end of the basin. This deepened basin helps keep the outflow zone free of emergent vegetation and makes the outlet less likely to clog. Wetlands serving small tributary watersheds will require small outlets to ensure that the drain time of the design capture volume is no fewer than 20 hours. However, designing small outlets that do not clog is difficult, if not impossible. For this case, a set of V-notch weirs or a sawtooth weir may be more appropriate for the outflow control device.

Vegetation. Suitable plants for created wetlands vary between different ecoregions. However, the wetland plants chosen for created wetlands should incorporate the following attributes:

- Tolerance to wide ranges of water elevations, salinity (salt content), temperature, and pH;
- A mixture of perennials and annuals;
- Moderate amounts of leaf production; and
- Proven removal efficiencies, for example, of *Scriptus* species.

Wetland plants are now commercially available in some municipalities from local nurseries who can provide additional information on tolerances and growth rates.

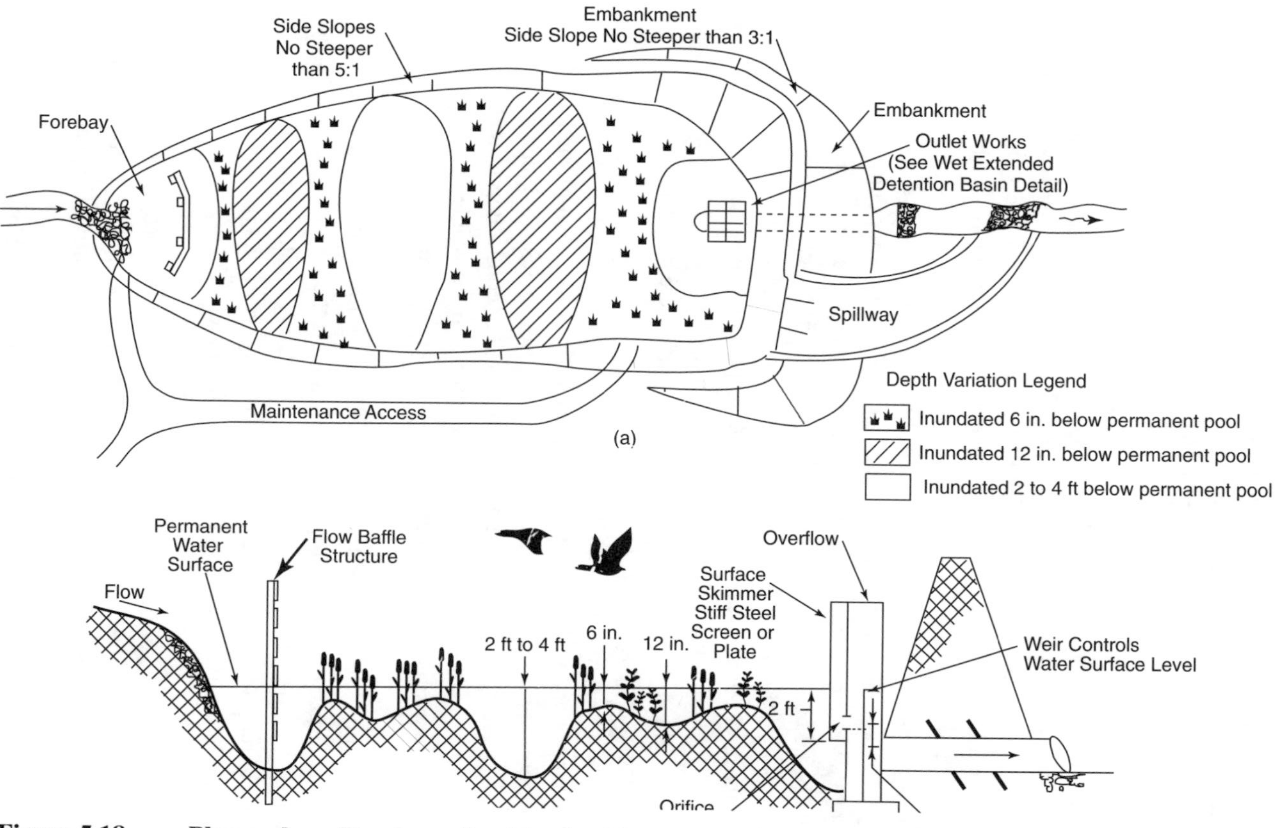

Figure 5.18 Plan and profile of a wetland basin: (a) plan (not to scale) and (b) profile (not to scale) (ft × 0.304 8 = m; in. × 25.4 = mm).

CONSTRUCTION. Construction management of the wetland is critical. Elevations and contouring of the constructed wetlands are the most important aspect of created wetlands—especially with respect to the groundwater. The confining layers of clay will vary from place to place within a specific area. If the clay layer is breached, clay should be replaced and the elevations of that location should be changed in the design.

Proper staging and sequencing will provide areas for dewatering during construction to reduce effects to adjacent waters. Rim ditches are particularly beneficial to avoid excessive pumping. When the created wetland is adjacent to an existing wetland, a temporary berm is needed until final grade has been achieved.

The use of organic soils is necessary to provide moisture-retaining abilities during drier periods and nutrients. If at all possible, soils (muck) from displaced wetlands should be stockpiled and used in the created wetland. Care should be taken to ensure wetland soils with nuisance species are not used. The displaced muck will provide root propagules, seed sources, micro- and mieofauna, and other invertebrates. Topsoil or peat can be substituted in the place of wetland muck. The muck/organic layer should be 0.1 to 0.3 m (6 to 12 in.) deep. Depths greater than 0.3 m (12 in.) tend to create difficulties in spreading the muck and planting.

If possible, it is best to control the hydration of the newly constructed wetland. The installation of plant material is most efficiently accomplished in saturated conditions, but standing water can cause poorly installed plants to float. Flash-board risers and adjustable gates can aid in controlling water levels during construction. However, if water control is not possible, wetland plants should be acclimated to inundation in the nursery before shipping to the site.

Keeping the soils saturated for 1 week after the muck/mulch has been spread will encourage seeds and propagules to sprout. If the wetland creation area is flooded with 0.15 m (6 in.) of water, by the second week this flooding will selectively remove upland species. The remaining water can be allowed to fill to design level after 3 weeks (Tesket and Hinckley, 1977).

MONITORING. Monitoring the created wetland will ensure proper coverage of the planted zones by desirable species. Monitoring should be done quarterly for the first year, semiannually for the second and third years, and (when necessary) annually for the fourth and fifth years.

Monitoring the wetland for the following information will help prevent future problems:

- Percent survivorship of planted species—subsamples can be used to provide quantitative results in larger wetlands,
- Percent cover of planted species and recruited desirable plants,
- Percent cover of nuisance species,
- Wildlife use, and
- Qualitative assessments of water quality.

Replanting as necessary to achieve the 85% survival rate at the end of each year is beneficial. It is critical to assess the created wetland for nuisance species. If the nuisance coverage is greater than 10%, maintenance through removal may be necessary.

MAINTENANCE. Maintenance includes three primary areas: replanting, nuisance species removal, and excavation of sediment sumps. Adjustments in plant type may be needed to accommodate differences in elevations. Typically, this is easier than regrading an established area. If water levels are lower than desired, adjustment of the control structure can increase survival.

Removal of nuisance species will increase the function and value of the constructed wetland. Harvesting wetland plants can be considered for nutrient removal but can resuspend trapped sediments. This resuspension, with habitat disturbance, is probably less beneficial than the actual nutrient removal.

MEDIA FILTRATION

Figure 5.19 shows a conceptual rendering of a media filtration facility widely used in the U.S. city of Austin, Texas. It consists of a settling basin followed by a filter. Field research indicates the sand filter has a suspended solids removal efficiency similar to that of retention ponds and extended detention.

The most typical filter is sand, but some use a peat and sand mixture because peat has the adsorptive ability to remove organics and dissolved contaminants. Clogging problems have been reported with the peat mixture (Tomasak *et al.,* 1987), but this may be because of the type of peat used (Galli, 1990). Limited research also indicates that compost made from leaves can be effective at removing dissolved phosphorus, metals, oil, and grease (Stewart, 1989), but field data show inconsistent performance.

Presettlement is essential to avoid rapid clogging of the filter. The device should not be on line during construction in the tributary watershed. Filters should be designed to have overflow or bypass for extreme events to protect against flooding because of backups if filter plugging occurs. It should be noted, however, that providing the bypass allows a clogged filter to appear to be operating effectively. The only way to ensure that the filter is not plugged is faithful periodic inspection providing a design that clearly indicates when the filter is no longer draining, such as excessive surface ponding or excessive bypasses by runoff.

THE AUSTIN, TEXAS, FILTER. The most extensive experience to date is with surface facilities of the type shown conceptually in Figure 5.19. This type of facility has been used on catchments of up to 20 ha (50 ac) in Austin, Texas, where it originated. Austin provides for two designs—one with full sedimentation and the other with partial sedimentation.

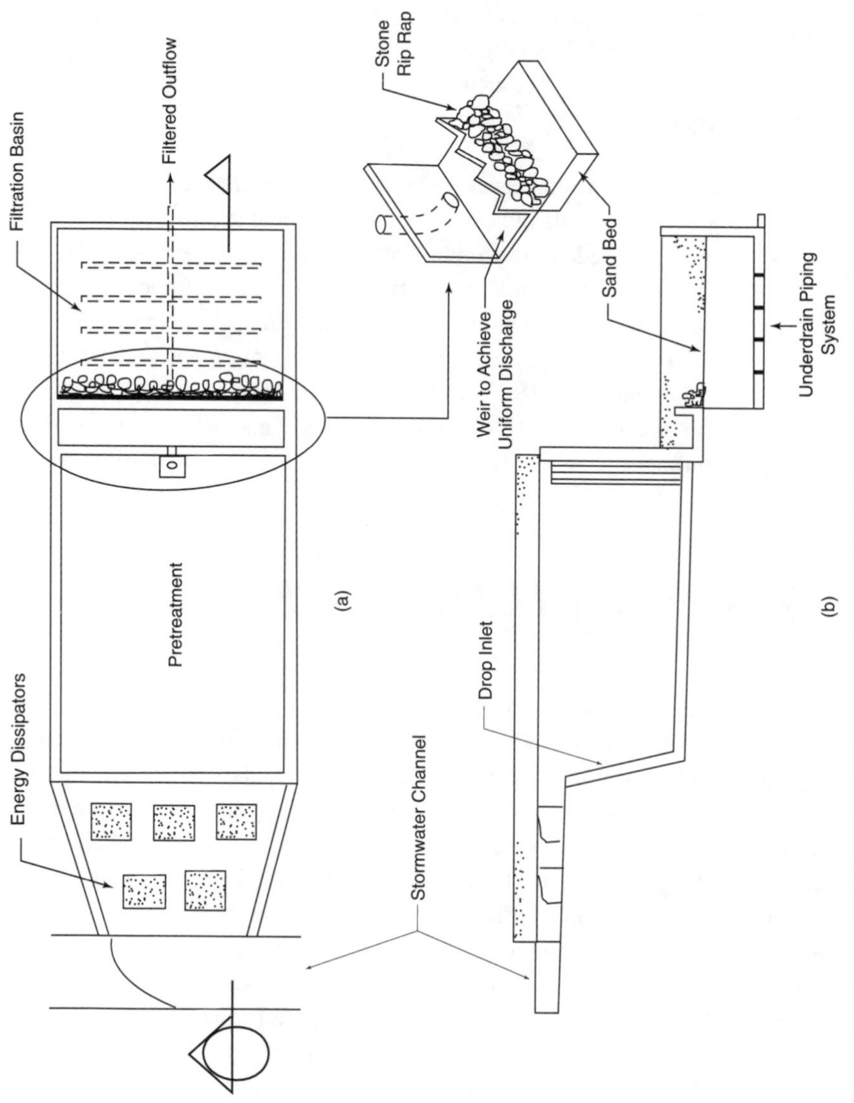

Figure 5.19 Sand filter with a presettlement basin: (a) plan view and (b) elevation.

The full sedimentation configuration includes a sedimentation basin designed to hold the entire water quality volume (that is, equivalent to the 40-hour drain time maximized volume) and to release this volume to the filter over a 40-hour drawdown period. This system should be used unless topographical constraints make this design unfeasible.

The partial sedimentation configuration requires less depth than the full sedimentation system and may be applicable where topographical constraints exist. In this system, a smaller sedimentation chamber is located upstream of the filtration basin and is designed to remove the heavier sediment and trash litter only. It requires more intensive maintenance than the full sedimentation system. The volume of the sediment chamber should be no less than 20% of the water quality volume used for the full sedimentation design. The design must ensure that the sediment chamber discharges the flow evenly to the filtration basin. Rock gabions composed of 300- to 450-mm (12- to 18-in.) diameter rocks can be used for this purpose. The outflow side of the sediment chamber should incorporate features to prevent gouging of the sand media (for example, concrete splash pad or riprap).

Determining the Surface Area of the Filter. This procedure is based on the design guidance provided in Austin, Texas. The design provides that a design water quality volume, namely the maximized volume, is processed through the filter without overtopping or bypassing the facility. The filter is sized as follows:

- Average hydraulic head on the filter is approximately 0.9 m (3 ft).
- For the full sedimentation design: 40-hour drawdown rate of the water quality volume and coefficient of permeability K = 1.1 m/d (3.5 ft/d).
- For the partial sedimentation design: 8-hour drawdown rate of the water quality volume and coefficient of permeability K = 0.6 m/d (2 ft/d).
- Clean concrete aggregate sand 0.02 to 0.04 in. in diameter (ASHTO C-33).

The above criteria result in the following equation:

$$A_F = \frac{A_T \cdot P_O}{12 \cdot K \cdot T_D} \tag{5.13}$$

Where

A_F = filter area, m^2 (sq ft);
A_T = area of the tributary catchment, m^2 (sq ft);
P_o = runoff volume equal to the maximized volume, mm (in.);
K = coefficient of permeability, m/d (ft/d); and
T_D = drawdown time of the maximized volume (40 hours for full sedimentation and 8 hours for partial sedimentation).

As an example, assume a 0.37-ha (40 000-sq ft) commercial catchment in the U.S. city of Dallas, Texas, with a runoff coefficient C = 0.70. Size a filter for the full sedimentation and the partial sedimentation conditions. Using Equation 5.2, the maximized runoff volume for this catchment is

P_O = 18.5 mm (0.73 in.) for the full sedimentation condition; and

P_O = 13.2 mm (0.55 in.) for the partial sedimentation condition.

Thus, for the full sedimentation condition, Equation 5.13 results in

$$A_F = \frac{(40\ 000)(0.73)}{12 \cdot (3.5) \cdot (40/24)} = 39 \text{ m}^2 \text{ (417 sq ft)}$$

and for the partial sedimentation condition, it becomes

$$A_F = \frac{(40\ 000)(0.55)}{12 \cdot (2.0) \cdot (8/24)} = 255 \text{ m}^2 \text{ (2 750 sq ft)}$$

Both filters will pass the maximized storm runoff volume or smaller without overtopping or bypassing any runoff, provided the filter is not excessively clogged. However, the full sedimentation design will have an extended detention basin upstream of the filter equal to 18.5 mm (0.73 watershed in.) of runoff, while the partial sedimentation design will have a basin with a volume equal to 20% of this volume. The design of choice will be governed by site constraints, maintenance considerations, and the owner's choices of which arrangement best fits the commercial site.

Configuring a Sand Filter. In addition to the coarse sediment and trash interception provided by the upstream detention storage, consider the following:

- Provide a trash rack at the outlet of the sedimentation basin,
- Provide access for maintenance equipment,
- Provide freeboard or a safe bypass when basin is full,
- Provide a sediment trap at the inlet to reduce resuspension,
- Use a flow spreader,
- Use one of two alternative sand bend designs (see Figure 5.20),
- Use a minimum sand bed thickness of 0.4 m (18 in.), and
- Provide underdrains under the sand.

LINEAR FILTERS—DELAWARE. An underground "linear" filter (see Figure 5.21) used in Delaware (U.S.) is suggested by Shaver (1991) for catchments of up to 2 ha (5 ac). This underground system uses a vault with a permanent pool of water as the pretreatment device. Shaver (1991) recommends that the volume of both the sedimentation and the filter chambers be approximately 38 m^3/ha (540 cu ft per contributing ac) and that the surface area of each chamber be 25 m^3/ha (360 sq ft per contributing ac). Configure the filter as follows:

- Depth of sand (0.4 m, or 18 in.).

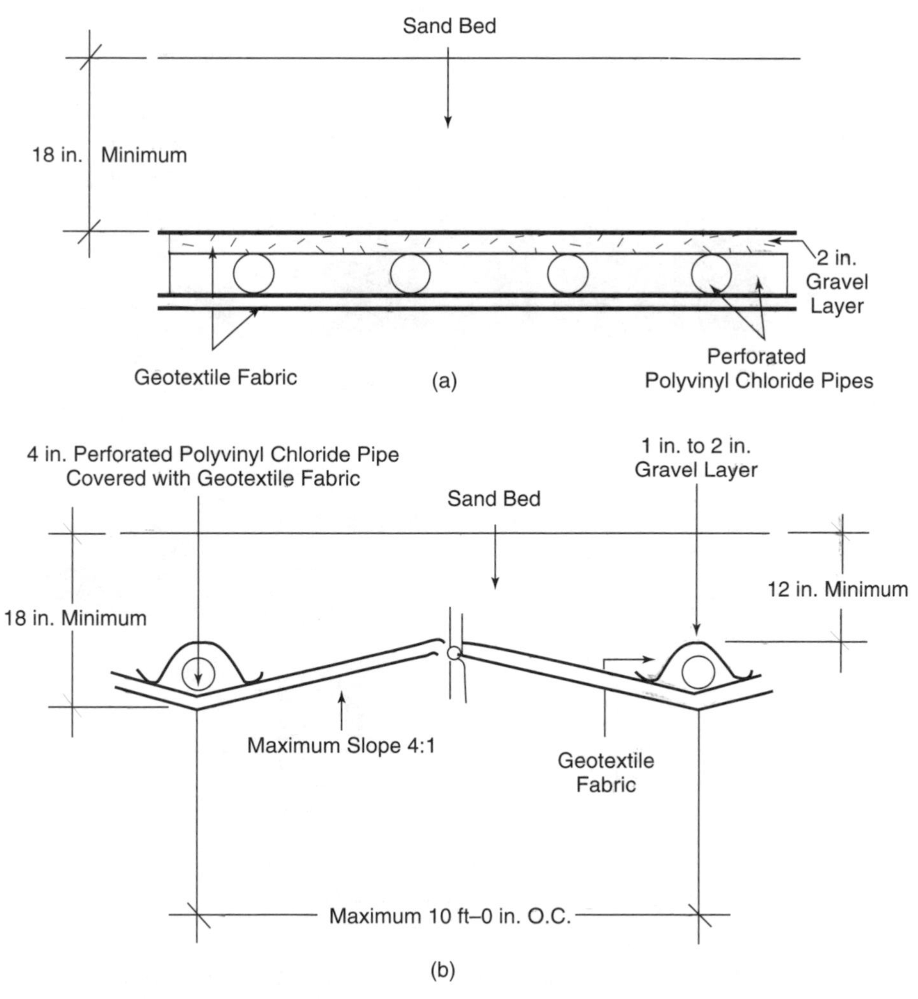

Figure 5.20 Sand bed filtration configurations: (a) sand bed profile (with gravel layer) and (b) sand bed profile (trench design) (ft × 0.304 8 = m; in. × 25.40 = mm) (City of Austin, 1989).

- Diameter of the outlet pipe should be 0.1 m (6 in.) or less (use multiple outlets if necessary).
- Position the filter relative to the pavement to evenly distribute the flow as it enters the sedimentation chamber. Pavement and inlet design and construction are, therefore, critical.

UNDERGROUND VAULT—WASHINGTON, D.C. A similar type of linear filter, illustrated in Figure 5.22, has been used in the Washington, D.C., area (U.S.). It is suggested that this particuiar design be viewed cautiously for the following reasons. First, the initial settling chamber is undersized for effective sedimentation, causing the filter to clog quickly. Second,

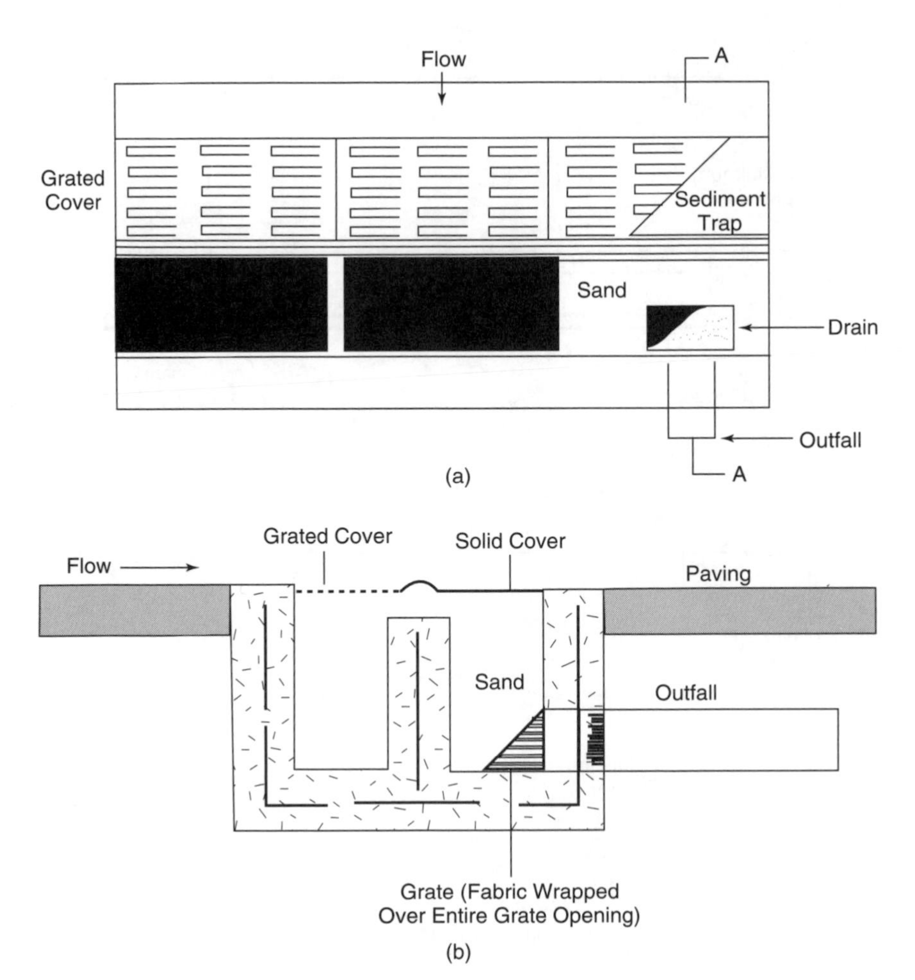

Figure 5.21 **Linear filter: (a) plan view and (b) section A-A (Shaver, 1995).**

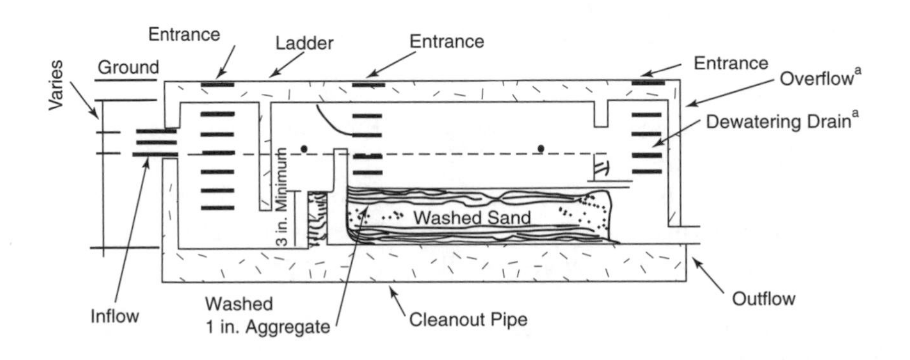

Figure 5.22 **Underground vault filter ([a]an ungated dewatering drain and overflow are not recommended) (in. × 25.40 = mm).**

when the filter clogs, the flow will simply overtop the overflow weir and flow directly to the outlet, with no indication that the filter is plugged. If this filter type is to be used, it should be sized using the Delaware linear filter criteria, including the presettlement chamber. It is also strongly recommended that the overflow weir and dewatering drain in the filter chamber be blocked and that the entrance manhole covers over the sedimentation chamber and the outflow chamber be replaced with grates. Then, if the filter clogs, the water will back up in the vault, overflow out of the inlet grate over the sedimentation compartment, and back into the outfall chamber. This will be a clear visual indication that the filter is plugged.

MAINTENANCE. Inspect semiannually and after major storms. Sediment and all floatables should be removed: from the settling basin when 100 mm (4 in.) accumulates; from the filter when 2.5 mm (0.1 in.) or more accumulates; or when there is standing water over the filter 40 hours after the storm. Field experience in Austin, Texas (U.S.), indicates the filter surfaces must be cleaned about twice each year by raking off the dried sediment. If there are open space areas in the tributary catchment that are erosive, or if construction is occurring, more frequent cleaning will be necessary. Consult Shaver (1991) and Truong (1989) for additional design and maintenance criteria.

OIL AND WATER SEPARATORS

Oil and water separators are designed to remove petroleum compounds, grease, and grit. They will also remove other floatable debris. Two types of oil and water separators—conventional gravity separators and the coalescing plate interceptor (CPI)—are used at all bulk petroleum storage and refinery facilities. The lack of oil characteristics and the apparently low concentrations of oil in most stormwater result in considerable performance uncertainty.

APPLICATION. This treatment control is applicable when the concentrations of oil- and grease-related compounds are abnormally high and source control does not provide effective control. The typical business types of concern are gasoline stations and truck, car, and equipment maintenance and washing enterprises and other commercial and industrial facilities that generate high levels of oil products in runoff wastes. Public facilities for which separators may be considered include marine ports, airfields, fleet vehicle maintenance and washing facilities, and mass transit park-and-ride lots.

PERFORMANCE. Conventional separators are capable of removing oil droplets with diameters equal to or greater than 150 microns. A CPI separator should be used if smaller droplets must be removed. When the droplet size is of sufficient size, oil and grease concentrations can be reduced to 10 mg/L or less (Lettenmaier and Richey, 1985).

Separator sizing is based on the rise velocity of an oil droplet, using oil density and droplet size to calculate rise velocity or using direct measurement of rise velocities. With the exception of stormwater from oil refineries, there are no relevant design data describing the characteristics of petroleum products in urban stormwater. A portion of the petroleum products are attached to fine suspended solids and, therefore, are removed by settling not flotation. Consequently, the performance of oil–water separators for urban stormwater runoff is uncertain.

DESIGN. The basic configurations of the two types of separators are illustrated in Figure 5.23. With small installations, a conventional gravity separator has the general appearance of a septic tank but is longer with respect to its width. Larger facilities have the appearance of a municipal wastewater primary sedimentation tank. The CPI separator contains closely spaced plates that enhance the removal efficiency and consequently requires less space than a conventional separator. The angle of the plates to the horizontal ranges from 0 (horizontal) to 60 deg, although 45 to 60 deg is typical. The perpendicular distance between the plates typically ranges from 19 to 25 mm (0.75 to 1.0 in.). The stormwater will flow either across the plates or down through the plates, depending on the plate configuration.

SIZING. The sizing of a separator is based on the calculation of the rise velocity of the oil droplets using Equation 5.14 (modified from API, 1990):

$$V_P = \frac{5.76 \cdot (d_p - d_c) \cdot d^2}{n} \cdot 10^{-15} \qquad (5.14)$$

Where

V_p = rise velocity, m/s (ft/sec);
d_p = density of the oil, kg/m^3 (lb/cu ft);
d_c = density of the water, kg/m^3 (lb/cu ft);
d = diameter of the droplet to be removed, m (ft); and
n = absolute viscosity of the water, kg/m^2 (lb/sq ft).

An appropriate water temperature value for selecting water density and viscosity is the expected temperature of the stormwater during the winter period. There are no data on the specific gravity of petroleum products in urban stormwater, but values between 0.85 and 0.95 typically are used. Also, distribution of droplet sizes must be known to select the appropriate droplet diameter for a stated efficiency goal. However, there is little information on the size distribution of oil droplets in urban stormwater. An oil droplet size and

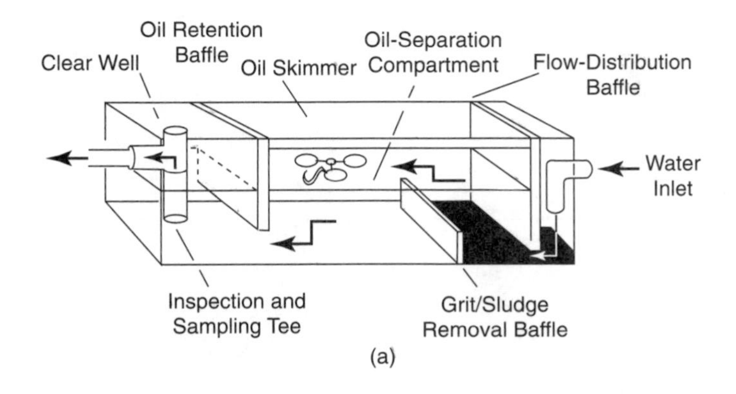

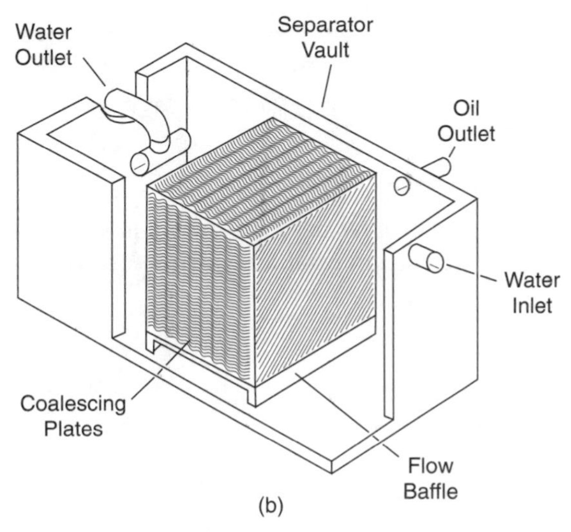

Figure 5.23 **(a) Conventional and (b) coalescing plate separators.**

volume distribution for stormwater from a petroleum products storage facility is depicted in Figure 5.24. Because a design influent concentration must be assumed, there will be considerable uncertainty because it will vary widely within and between storms.

If the effluent goal is 20 mg/L and the design influent concentration is 50 mg/L, a removal efficiency of 60% is required. Using Figure 5.24, this efficiency can be achieved by removing all droplets with diameters 90 microns or larger. Using a water temperature of 10°C (a water density of 0.999) and an oil density of 0.898, the rise velocity for a 90-micron droplet is 1.2 m/h (0.001 1 ft/sec).

It is typically believed that conventional separators are not effective at removing droplets smaller than 150 microns (API, 1990). Theoretically, a conventional separator can be sized to remove a smaller droplet, but the facility may be so large that the CPI separator may be more cost effective.

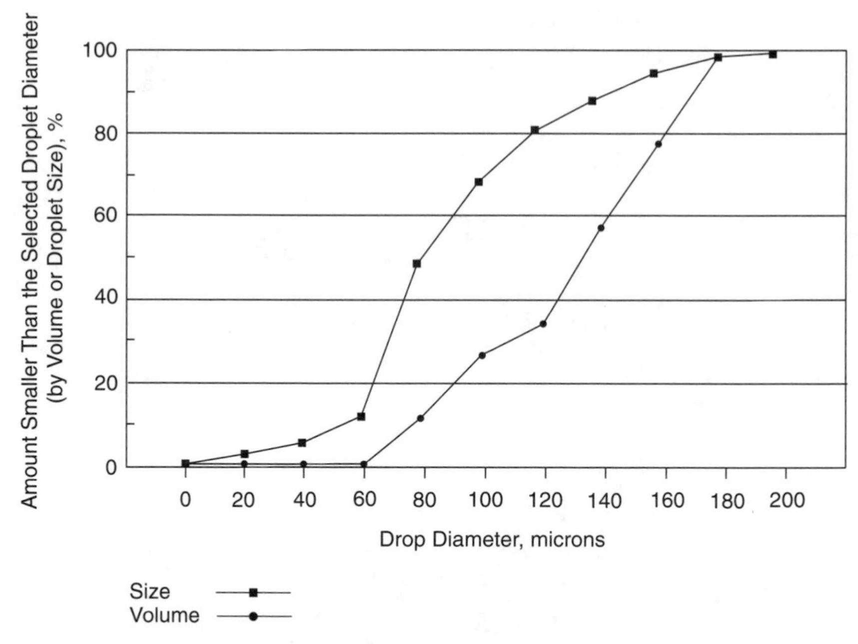

Figure 5.24 **Oii droplet size distribution in stormwater from petroleum products storage facilities.**

Sizing Conventional Separators. To size a conventional separator, first compute the depth:

$$D = \sqrt{\frac{Q}{2V}} \tag{5.15}$$

Where

D = depth, which should be between 0.9 and 2.4 m (3 and 8 ft);
Q = design flow rate, m^3/s (cfs); and
V = allowable horizontal velocity, no more than 15 times the design oil rise rate but not greater than 55 m/h (0.05 ft/sec).

If the computed depth exceeds 2.4 m (8 ft), design additional parallel units such that, at the design flow rate, the maximum recommended depth of 2.4 m (8 ft) is not exceeded. Minimum depth is 0.9 m (3 ft).

The next steps are

- Calculate length, $L = I \cdot D/V_p$;
- Select width, W = 2 to 3 times the depth, but not to exceed 6 m (20 ft);
- Baffle height-to-depth ratio of 0.85 for top baffles and 0.15 for bottom baffles;
- Locate the distribution baffle at $0.10L$ from the entrance;

- Add 0.3 m (1 ft) for freeboard; and
- Install an inlet flow control and a bypass for flows in excess of the design flow.

Sizing Separators. Manufacturers can provide packaged separator units for flows up to several cubic metres (cubic feet) per second. For larger flows, the engineer must size the plate pack and design the vault. Given the variability of separator technology among manufacturers with respect to plate size, spacing, and inclination, it is recommended that the design engineer consult vendors for a plate package that will meet the engineer's criteria.

The engineer can size the facility using the following procedure. First, identify the expected plate angle (vertical above horizontal), *H*, in degrees, and calculate the total plate area required, *A*, in square metres (square feet):

$$A = \frac{Q}{V_P \cdot \cos(H)} \tag{5.16}$$

where the terms are the same as defined under Equations 5.14 and 5.15.

Coalescing plate interceptor separators are not 100% hydraulically efficient, ranging from 0.35 to 0.95 depending on the plate design. If the engineer wishes to incorporate this factor, divide the result from Equation 5.16 by the selected efficiency.

- Select spacing, *S*, between the plates, typically 19 to 38 mm (0.75 to 1.5 in.).
- Identify reasonable plate width, *W*, and length, *L*.
- Number of plates $N = A/(W \cdot L)$.
- Calculate plate volume, P_v, m^3 (cu ft).

$$P_V = \left[\frac{N \cdot S}{12} + L \cdot \cos(H)\right] \cdot [W \cdot L \cdot \sin(H)] \tag{5.17}$$

- Add 0.3 m (1 ft) beneath the plates for sediment storage.
- Add 0.1 to 0.3 m (6 to 12 in.) above the plates for water clearance so that the oil accumulates above the plates.
- Add 0.3 m (12 in.) for freeboard.
- Add a forebay for floatables and distribution of flow if more than one plate unit is needed.
- Add an afterbay for collection of the effluent from the plate pack area.
- For larger units include a device to remove and store oil from the water surface.

Horizontal plates require the least plate volume to achieve a particular removal efficiency.

Settleable solids will accumulate on the plates, complicating maintenance procedures. Experience shows that, even with slanted plates, some solids will stick to the plates because of the oil and grease. If debris is expected such as sticks, plastics, and paper, then select a larger plate separation distance. As an alternative, install a trash rack or screens with smaller openings than the plate spacing. The plates may be damaged by the weight when removed for cleaning.

MAINTENANCE. Check monthly during the wet season and clean several times a year. Always clean before the start of the wet season. Properly dispose of the oil.

REFERENCES

American Petroleum Institute (1990) *Design and Operation of Oil-Water Separators.* Publication 421, Washington, D.C.

American Society of Civil Engineers (1985) Final Report of the Task Committee on Stormwater Detention Outlet Control Structures. Am. Soc. Civ. Eng., New York, N.Y.

American Society of Civil Engineers and Water Environment Federation (1992) *Design and Construction of Urban Stormwater Management Systems.* Am. Soc. Civ. Eng. Manuals and Reports of Engineering Practice No. 77, New York, N.Y.; Water Environ. Fed. Manual of Practice No. FD-20, Alexandria, Va.

Camp Dresser & McKee, *et al.* (1993) *California Storm Water Best Management Practices Handbooks.* Public Works Agency, County of Almeda, Calif.

City of Austin (1989) *Environmental Criteria Manual, Design Guidelines for Water Quality Control.* Austin, Tex.

DeGroot, W.G. (1982) *Stormwater Detention Facilities.* Am. Soc. Civ. Eng., New York, N.Y.

Driscoll, E.D. (1983) Performance of Detention Basins for Control of Urban Runoff Quality. *Proc. Int. Symp. Urban Hydrol., Hydraul. and Sediment Control,* Univ. Kentucky, Lexington.

Driscoll, E.D., *et al.* (1989) *Analysis of Storm Events, Characteristics for Selected Rainfall Gauges Throughout the United States.* U.S. EPA, Washington, D.C.

Florida Department of Environmental Regulation (1988) *The Florida Development Manual: A Guide to Sound Land and Water Management.* Fla. Dep. Environ. Resour., Nonpoint Source Manage. Sect., Tallahassee.

Galli, J. (1990) *Peat-Sand Filters: A Proposed Stormwater Management Practice for Urbanized Areas.* Metro. Wash. Council Gov., Washington, D.C.

Grizzard, T.J., *et al.* (1986) Effectiveness of Extended Detention Ponds. In *Urban Runoff Quality—Impact and Quality Enhancement Technology.* Am. Soc. Civ. Eng., New York, N.Y.

Guo, C.Y., and Urbonas, B.R. (1995) Special Report to the Urban Drainage and Flood Control District on Stormwater BMP Capture Volume Probabilities in United States. Denver, Colo.

Hartigan, J.P. (1989) Basis for Design of Wet Detention Basin BMPs. In *Design of Urban Runoff Quality Controls.* L.A. Roesner *et al.* (Eds.), Am. Soc. Civ. Eng., New York, N.Y.

Lettenmaier, D., and Richey, J. (1985) *Operational Assessment of a Coalescing Plate Oil/Water Separator.* Munic. Metro., Seattle, Wash.

Livingston, E.H. (1989) The Use of Wetlands for Urban Stormwater Management. In *Design of Urban Runoff Quality Controls.* Am. Soc. Civ. Eng., New York, N.Y.

Livingston, E.H., *et al.* (1988) *The Florida Development Manual. A Guide to Sound Land and Water Management.* Dep. Environ. Regulation, Tallahassee, Fla.

Mills, W.B., *et al.* (1982) *Water Quality Assessment: A Screening Procedure for Toxic and Conventional Pollutants.* EPA-600/6-82-004, U.S. EPA, Environ. Res. Lab., Athens, Ga.

Randall, C.W., *et al.* (1982) Urban Runoff Pollutant Removal by Sedimentation. In *Stormwater Detention Facilities.* Am. Soc. Civ. Eng., New York, N.Y.

Roesner, L.A., *et al.* (1989) Design of Urban Runoff Quality Controls. *Proc. Eng. Found. Conf. Curr. Pract. Des. Criteria Urban Qual. Cont.*, Am. Soc. Civ. Eng., New York, N.Y.

Roesner, L.A., *et al.* (1991) Hydrology of Urban Runoff Quality Management. *Proc. 18th Natl. Conf. Water Res. Plann. Manage., Symp. Urban Water Res.*, Am. Soc. Civ. Eng., New Orleans, La.

Schueler, T.R. (1987) *Controlling Urban Runoff: A Practical Manual for Planning and Designing Urban Best Management Practices.* Metro. Wash. Water Resour. Plann. Board, Washington D.C.

Schueler, T.R., *et al.* (1992) *Current Assessment of Urban Best Management Practices—Techniques for Reducing Non-Point Source Pollution in the Coastal Zone.* Metro. Council Gov., Washington, D.C.

Shaver, E. (1995) Sand Filter Design for Water Quality Treatment. In *Stormwater Runoff and Receiving Systems.* E.E. Herricks (Ed.), Lewis Publishing, New York, N.Y.

Stewart, W. (1989) Evaluation and Full-Scale Testing of a Compost Biofilter for Stormwater Runoff Treatment. Paper presented at Annu. Conf. Pacific Northwest Pollut. Cont. Assoc.

Strecker, E.W., *et al.* (1992) *The Use of Wetlands for Controlling Stormwater Pollution.* Terrene Inst., Washington, D.C.

Swedish Water and Sewage Works Association (1983) Local Disposal of Storm Water (Swed.). Publication VAV P46, Stockholm, Swed.

Tesket, R.O., and Hinckley, T.M. (1977) Impact of Water Level Changes on Woody Riparian and Wetland Communities, U.S. Fish and Wildlife Serv., PB-276 036, Washington, D.C.

Tomasak, M.D., *et al.* (1987) Operational Problems with a Soil Filtration System for Treating Stormwater. Minn. Pollut. Control Agency, St. Paul, Minn.

Truong, H.V. (1989) The Sand Filter Quality Structure. Dist. Columbia Gov., Washington, D.C.

U.S. Environmental Protection Agency (1983) *Results of the Nationwide Urban Runoff Program.* Final Report, Water Plann. Div., Washington, D.C.

U.S. Environmental Protection Agency (1986) *Methodology for Analysis of Detention Basins for Control of Urban Runoff Quality.* EPA-440/5-87-001, Washington, D.C.

Urban Drainage and Flood Control District (1992) *Urban Storm Drainage Criteria Manual: Volume 3—Best Management Practices.* Denver, Colo.

Urbonas, B.R. (1994) Technical Hint. *Flood Hazard News.* Urban Drainage and Flood Control Dist., Denver, Colo.

Urbonas, B.R., and Roesner, L.A. (Eds.) (1986) Urban Runoff Quality—Impact and Quality Enhancement Technology. *Proc. Eng. Found. Conf.*, Am. Soc. Civ. Eng., New York, N.Y.

Urbonas, B.R., and Stahre, P. (1993) *Stormwater—Best Management Practices Including Detention.* Prentice Hall, Englewood Cliffs, N.J.

Urbonas, B.R., *et al.* (1990) Optimization of Stormwater Quality Capture Volume. In *Urban Stormwater Quality Enhancement—Source Control, Retrofitting and Combined Sewer Technology.* Am. Soc. Civ. Eng., New York, N.Y.

Walker, W.W. (1985) Empirical Methods for Predicting Eutrophication in Impoundments—Report 3: Model Refinements. Tech. Rep. E-81-9, U.S. Army Corps Eng., Waterways Exp. Stn., Vicksburg, Miss.

Walker, W.W. (1987) Phosphorus Removal by Urban Runoff Detention Basins. Lake and Reservoir Management: Volume III. North Am. Lake Manage. Soc., Washington, D.C.

Whipple, W., and Hunter, J.V. (1981) Settleability of Urban Runoff Pollution. *J. Water Pollut. Control Fed.*, **53,** 1726.

Wiegand, C., *et al.* (1989) Cost of Urban Quality Controls. In *Design of Urban Runoff Quality Controls.* Am. Soc. Civ. Eng., New York, N.Y.

Index

A

B

C

D

E

G

H

I

L

M

N

O

P

R

S

T

U